D0205457

Elementary Probability
with Applications

Elementary Probability
with Applications

Larry Rabinowitz

A K Peters
Wellesley, Massachusetts

Editorial, Sales, and Customer Service Office

A K Peters, Ltd.
888 Worcester Street, Suite 230
Wellesley, MA 02482
www.akpeters.com

Cover Illustration: Random number target, see Figure 1.1 on page 6.

Library of Congress Cataloging-in-Publication Data

Rabinowitz, Larry, 1937–
     Elementary probability with applications / Larry Rabinowitz.
          p. cm.
     Includes bibliographical references and index.
     ISBN 1-56881-222-1
     1. Probabilities. I. Title.

QA273.R16 2004
519.2--dc22   2004051556

Printed in Canada
09 08 07 06 05   10 9 8 7 6 5 4 3 2 1

"In the nineteenth century, mathematicians attacked pi with the help of human computers. The most powerful of these was Johann Martin Zacharias Dase, a prodigy from Hamburg. Dase could multiply large numbers in his head, and he made a living exhibiting himself to crowds in Germany, Denmark, and England, and hiring himself out to mathematicians. A mathematician once asked Dase to multiply 79,532,853 by 93,758,479, and Dase gave the right answer in fifty-four seconds. Dase extracted the square root of a hundred-digit number in fifty-two minutes, and he was able to multiply a couple of hundred-digit numbers in his head during a period of eight and three-quarters hours. Dase could do this kind of thing for weeks on end, running as an unattended supercomputer. He would break off a calculation at bedtime, store everything in his memory for the night, and resume calculation in the morning. Occasionally, Dase had a system crash. In 1845, he bombed while trying to demonstrate his powers to a mathematician and astronomer named Heinrich Christian Schumacher, reckoning wrongly every multiplication that he attempted. He explained to Schumacher that he had a headache. Schumacher also noted that Dase did not in the least understand theoretical mathematics. A mathematician named Julius Petersen once tried in vain for six weeks to teach Dase the rudiments of Euclidean geometry, but they absolutely baffled Dase. Large numbers Dase could handle, and in 1844 L. K. Schulz von Strassnitsky hired him to compute pi. Dase ran the job for almost two months in his brain, and at the end of the time he wrote down pi correctly to the first two hundred decimal places—then a world record."

from "The Mountains of Pi" by Richard Preston
The New Yorker, 1992

Contents

Preface

This book is based on a set of notes I have used to teach an elementary probability course at The College of William and Mary. The book is designed for either a one semester or a one quarter course in discrete probability where there is a strong emphasis on applications. Except for mathematics, computer science, chemistry, and physics, the students come from all departments of the College. No previous knowledge of probability is assumed and in fact the material can be read by someone who has taken only a first course in high school algebra. At the same time, some of the reading material is challenging because the applications are intricate. In addition, some exercises are difficult and the student should be prepared to spend time on them. I usually assign an average of 3 or 4 homework problems each class for a three credit course that meets for fifty minutes three times a week. I try to stagger the exercises with respect to difficulty. So one assignment might consist of a difficult problem from the previously studied section, and two or three less difficult problems from a new section. Some chapter exercises can be the basis for a class lecture.

By studying this book the student will learn to appreciate the subject of probability and its applications and will develop his or her problem-solving and reasoning skills. Probability has practical uses in many different fields. The focus is on how probability models with underlying assumptions can be used to model real world situations.

The approach is informal. At this level, students learn from examples rather than theory. While the book contains some standard problems about dice and drawing chips from urns, the main focus is on real-world applications. My goal was to emphasize these applications rather than present a wide range of topics in probability. I tried to present only those concepts of probability needed to understand these applications. As a result, some topics which are usually presented in a book at this level are omitted here.

Many of the exercises in this book are based on examples from the chapter but are in a new setting. Others incorporate material from previous chapters. Students should not always expect to be able to write out the solution to a problem immediately after reading the question. This is not a "plug and chug" course. It is essential that students read the problem carefully to understand what the question is asking before attempting to find the solution. The exercises at the end of each chapter are roughly arranged in the same order as the material in that chapter.

The starred sections are optional and the subsequent material does not depend on them. The level of difficulty of the starred sections is comparable to that of the other sections. Starred problems are based on material from starred sections. All of Chapter 7 is optional.

Short answers to selected exercises can be found at the end of the book immediately preceding the bibliography. A calculator to carry out basic arithmetic operations such as addition, multiplication, division, and exponentiation will be needed to solve the exercises in the book.

1

Basic Concepts in Probability

1.1 Sample Spaces, Events, and Probabilities

A *sample space*, S, is defined as the set of all possible outcomes of an experiment. If the outcomes in S are equally likely, we call S an *equally probable sample space*. Any subset of S is called an *event*. For example, if a box contains three chips numbered 1,2,3 and we draw one chip from the box in such a way that each of the three chips are equally likely to be selected, then {1,2,3} is an equally probable sample space and ϕ(empty set), {1}, {2}, {3}, {1,2}, {1,3}, {2,3}, {1,2,3} are events. For this experiment {odd, even} is also a sample space but it is not an equally probable sample space since an odd outcome is twice as likely as an even outcome.

Let $P(A)$ denote the probability that event A will occur. If S is an equally probable sample space, then

$$P(A) = \frac{\text{number of outcomes in } A}{\text{number of outcomes in } S}.$$

Example 1.1. Box I contains chips numbered 1,2,3 and box II contains chips numbered 2,3,4. One chip is selected at random from each box. When we say "at random" we mean equally likely or in this example an equally likely selection.

Then $S = \{(1,2), (1,3), (1,4), (2,2), (2,3), (2,4), (3,2), (3,3), (3,4)\}$,

where the first number in each ordered pair denotes the number on the chip drawn from box I and the second number in each ordered pair denotes

1

the number on the chip drawn from box II. This is the set of all possible outcomes for this experiment. Since we have an equally likely selection from each box, the nine outcomes in S are equally likely and S is an equally probable sample space. Consider the event A where $A =$ "sum of the digits on the selected chips is an even number."

$$\text{Then } A = \{(1,3),(2,2),(2,4),(3,3)\}.$$

These are the only outcomes in S where the sum of the two digits is an even number. Since there are 4 outcomes in A,

$$P(A) = \frac{\text{number of outcomes in } A}{\text{number of outcomes in } S} = \frac{4}{9}.$$

What this means is that if we were to perform this experiment repeatedly over a long period of time, the proportion of times that we would get an even sum occurring will approach $\frac{4}{9}$ in the long run. It does *not* mean that if we were to repeat the experiment of drawing one chip at random from each box exactly nine times, we would get an even sum exactly four times. The point is that we should think of probability as a long-run proportion.

Another important distinction which needs to be made is that there is a difference between an event, A, and the probability of that event, $P(A)$. The event, A, is a set while $P(A)$ is a number. In Example 1.1, $A = \{(1,3),(2,2),(2,4),(3,3)\}$ whereas $P(A) = \frac{4}{9}$.

Example 1.2. Two red blocks and two green blocks are arranged at random (all possible arrangements are equally likely) in a row. Let $A =$ "there is exactly one green block between the two red blocks." Find $P(A)$.

Solution: To find $P(A)$ we first write out S.

$$S = \{(RRGG),(RGRG),(RGGR),(GGRR),(GRGR),(GRRG)\}.$$

Since all possible arrangements are equally likely, this is an equally probable sample space. Next list the outcomes in A. This means finding those outcomes in S where there is exactly one green block between two red blocks.

$$A = \{(RGRG),(GRGR)\}.$$

Thus,

$$P(A) = \frac{\text{number of outcomes in } A}{\text{number of outcomes in } S} = \frac{2}{6} = .33.$$

When listing the outcomes in S, try to do so in an organized manner. By doing this, you will have less of a tendency to omit outcomes or to list outcomes twice. What I attempted to do here in listing the outcomes in S was first to list those which begin with R and then move the second R throughout the other positions. Next begin with G and then move the other G throughout the other positions. Obviously, for large sample spaces an organized listing will be difficult. Many problems will have very large sample spaces where it would not be feasible to list all the outcomes in S. We will develop other techniques for handling such problems. But for now we will focus on reading and interpreting problems with small sample spaces.

Example 1.3. On the day the incoming freshmen arrive on campus someone has taken down the signs for three dormitories on campus and the freshmen are confused. To make matters worse, the person who replaced the signs wasn't sure about which sign went where but put the three signs up at random. What is the probability that none of the signs are correctly replaced?

Solution: We list the six possible outcomes in S.

Dormitory	Chandler	Barrett	Jefferson	
Dormitory signs	C	B	J	
	C	J	B	
	B	J	C	x
	B	C	J	
	J	B	C	
	J	C	B	x

In two of the six equally likely possibilities which are marked with an x, none of the signs are correctly replaced.
So,

$$P(\text{none of the signs are correctly replaced}) = \frac{2}{6} = .33.$$

The next example involves sampling without replacement.

Example 1.4. A box contains four chips numbered 1,2,3,4. Two chips are drawn, at random, without replacement, from the box. Let $A =$ "largest number selected is 3." Find $P(A)$.

Solution: Since we are drawing without replacement, we cannot get the same chip twice. So,

$$S \;=\; \{(1,2),(2,1),(1,3),(3,1),(1,4),(4,1),$$
$$(2,3),(3,2),(2,4),(4,2),(3,4),(4,3)\},$$

where the first entry in each ordered pair is the number on the first chip selected and the second entry in each ordered pair is the number on the second chip selected. Also,

$$A = \{(1,3),(3,1),(2,3),(3,2)\}, \quad \text{giving } P(A) = \frac{4}{12} = .33.$$

Note that corresponding to each outcome in S there is another outcome in S with the digits reversed. So, (1,3) and (3,1) are both included in S. A problem like this which involves sampling without replacement can also be solved by looking at outcomes which consider only content and not order. In other words, the two outcomes (1,3) and (3,1) could be replaced by one outcome which is an unordered pair containing 1 and 3. If we were to use this approach in Example 1.4 there would be one half as many outcomes in S and one half as many outcomes in A resulting in the same answer for $P(A)$. While this solution is correct, my suggestion is that we solve such problems listing ordered outcomes. If we had been sampling *with replacement*, it would have been necessary to consider ordered outcomes to arrive at a correct solution. (See Problem 15 in Chapter 1.) So by looking at ordered outcomes regardless of whether sampling is with or without replacement, we are using a procedure which will give us a correct solution in both cases without introducing much additional work so long as S is small.

We have a method for determining the probability of an event when the sample space is equally probable, but how should we proceed if the sample space is not equally probable? Let's look at an example for ideas.

Example 1.5. Let's go back to game four of the 1984 National Basketball Association championship series between the Celtics and the Lakers. With the score tied and only 35 seconds left in overtime, Earvin "Where Has the Magic Gone" Johnson steps to the foul line for two shots. He misses them both! Based on his previous record, he had been an 80% foul shooter. What is the probability that he missed those two shots due to chance? Do you think that "Magic" choked?

Solution: Consider the sample space $S = \{(H, H), (H, M), (M, H), (M, M)\}$, where H denotes hitting a shot, M denotes missing a shot. The first entry in each ordered pair indicates the outcome of the first shot while the second element of each ordered pair indicates the outcome of the second shot. So (M, H) represents missing the first shot and hitting the second shot. It is tempting to assume equally likely outcomes when this assumption is inappropriate. This is not an equally probable sample space since (H, H) is more likely than (M, M) as he is more likely to hit than to miss each shot. Since he is an 80% foul shooter we introduce the symbols: $H1, H2, H3, H4, M$ as the five equally likely cases for each shot. We have four H's because we want the probability that he hits a shot to be $\frac{4}{5} = 80\%$.

Thus an equally probable sample space for this experiment is:

$$S = \{(H1, H1), (H1, H2), (H1, H3), (H1, H4), (H1, M), (H2, H1),$$
$$(H2, H2), (H2, H3), (H2, H4), (H2, M), (H3, H1), (H3, H2),$$
$$(H3, H3), (H3, H4), (H3, M), (H4, H1), (H4, H2), (H4, H3),$$
$$(H4, H4), (H4, M), (M, H1), (M, H2), (M, H3), (M, H4), (M, M)\}.$$

The letter in the first entry of each ordered pair denotes whether he hits or misses on the first shot and the letter in the second entry indicates whether he hits or misses on the second shot. If $A = $ "misses both shots" $= \{(M, M)\}$, then, since there are 25 outcomes in S, $P(A) = \frac{1}{25} = .04$. The probability of an 80% foul shooter missing two foul shots by chance alone is .04. In other words, it is highly unlikely that he would miss both shots and we conclude that "Magic" choked.

It is important to make the four H's distinguishable in order to insure that each possible outcome is included exactly once in S. How we make the H's distinguishable such as by using $H1$ or H_1 is to some degree a matter of taste but we should try to keep the notation simple.

While we were able to solve Example 1.5 by listing outcomes repeatedly to get the appropriate weights so that we would have an equally probable sample space, it is clear that this technique would only be feasible in very simple cases. Later on, we will see how to solve problems of this type using some rules about probabilities.

1.2   Simulations

There are situations where either there is not an analytical solution or
we don't know how to find the analytical solution to a probability problem. In these cases we can estimate $P(A)$ by the long-run proportion of
times that A occurs. In other words, we simulate the experiment. For
this book, we will use a random number target (Figure 1.1) to carry out
these simulations. By closing one's eyes and bringing a pencil point down
on the target we are carrying out an experiment which is equivalent to
rolling a balanced die. That is, each of the numbers 1, 2, 3, 4, 5, 6 are
equally likely.

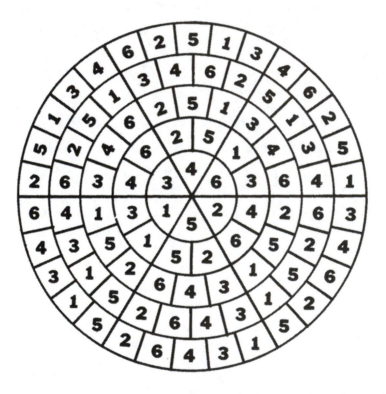

Figure 1.1. Random Number Target.

To demonstrate the simulation procedure we will estimate $P(A)$ for Example 1.1. Since we have already found $P(A)$ analytically, the sole purpose of estimating $P(A)$ here is to demonstrate the simulation procedure. Recall that in Example 1.1, we had box I containing chips numbered 1, 2, 3, and box II containing chips numbered 2, 3, 4. We select one chip at random from each box and want to estimate the probability that the sum of the numbers on the two chips drawn is even. Let's simulate this experiment 20 times. For the drawing from box I, we select a number from the random number target until we get either a 1, 2, or 3. So for this drawing, 4's, 5's, and 6's will be ignored. For the drawing from box II, we select numbers from the random number target until we get either a 2, 3, or 4. In this case 1, 5, and 6 are ignored.

In the following listing for my 20 simulations the first number in the ordered pair corresponds to the drawing from box I and the second number in the ordered pair corresponds to the drawing from box II.

Simulation Number			Simulation Number		
1:	(3,2)		11:	(2,4)	x
2:	(1,4)		12:	(3,3)	x
3:	(3,3)	x	13:	(2,3)	
4:	(3,2)		14:	(2,2)	x
5:	(2,4)	x	15:	(1,4)	
6:	(1,3)	x	16:	(3,2)	
7:	(3,4)		17:	(1,3)	x
8:	(2,3)		18:	(2,2)	x
9:	(3,2)		19:	(1,2)	
10:	(1,2)		20:	(3,4)	

We observe that the sum is even for the eight simulations which have been marked by an x. The estimate of $P(A)$ is the proportion of times which A occurs which is $\frac{8}{20}$ or .4. This is not $P(A)$. We have already evaluated $P(A)$ to be $\frac{4}{9}$. The value of .4 is an estimate of $P(A)$ based on the simulations. If each student in the class simulated this experiment 20 times we would get a number of different answers for the estimate of $P(A)$. Most of the estimates would be near $\frac{4}{9}$.

Note: The random number target contains 6 equally likely numbers. Suppose we were doing a simulation for a problem with more than 6 equally likely outcomes. We might have a problem with 9 equally likely outcomes.

One way of carrying out the simulation would be to take a pair of numbers from the random number target. There are $6 \times 6 = 36$ equally likely ordered pairs of numbers. One scenario would be to let four ordered pairs correspond to one outcome ($36/9 = 4$). Say,

(1,1)(1,2)(1,3)(1,4) corresponds to outcome #1, and

(1,5)(1,6)(2,1)(2,2) corresponds to outcome #2, etc.

Of course, we could have let one ordered pair correspond to one outcome. Say,

(1,1) corresponds to outcome #1, and

(1,2) corresponds to outcome #2, etc.

This method would use up only 9 of the 36 ordered pairs which means that the other 27 ordered pairs would have to be ignored leading to the selection of an average of four times as many random number pairs as needed.

Problem 9 at the end of this chapter will ask you how to estimate $P(A)$ for Example 1.1, without ignoring any numbers from the random number target.

Chapter exercises request a small number of simulations for simplicity. But in practice, a large number of simulations are required to obtain good estimates.

1.3   Complementary Events and Mutually Exclusive Events

Events A, B are said to be *mutually exclusive* if they cannot occur together. This means that the sets are disjoint (do not overlap). Events A, B are said to be *complementary* if both of the following conditions are satisfied:

(i) Events A, B are mutually exclusive.

(ii) Events A, B exhaust the sample space in the sense that every outcome in S is either in A or in B.

So, events "A," "not A" are complementary events. Equivalently, A and B are complementary events if $B =$ not A.

Example 1.6. Roll a fair die once. Let

$$\begin{aligned} A &= \text{``outcome is less than 5''} \\ B &= \text{``outcome is greater than 3''} \\ C &= \text{``outcome is less than 2''} \\ D &= \text{``outcome is greater than 4''} \end{aligned}$$

 (i) Which pair(s) of events are mutually exclusive?

 (ii) Which pair(s) of events exhaust S?

 (iii) Which pair(s) of events are complementary?

Solution: $S = \{1, 2, 3, 4, 5, 6\}$.

$$A = \{1, 2, 3, 4\}, \quad B = \{4, 5, 6\}, \quad C = \{1\}, \quad D = \{5, 6\}.$$

 (i) A, D are mutually exclusive
 B, C are mutually exclusive
 C, D are mutually exclusive

 (ii) A, B exhaust S
 A, D exhaust S

 (iii) From (i) and (ii), the only pair of events which are mutually exclusive and exhaust S are A, D. We observe that $A =$ not D.

Example 1.7. Suppose that singer/student Meat Loaf takes 5 courses for each of the 8 semesters that he is working on his college degree. Describe the complementary event to each of the following events:

 (i) $G =$ "At least one semester Meat Loaf gets all A's,"

 (ii) $H =$ "Meat Loaf makes at least one A every semester."

Solution:

 (i) This means that Meat Loaf gets 5 A's in either 1, 2, 3, 4, 5, 6, 7, or 8 semesters. The complementary event is that he makes 5 A's zero times. So, not $G =$ "he makes less than 5 A's every semester."

 (ii) This means that Meat Loaf makes either 1, 2, 3, 4, or 5 A's for all eight semesters. The complementary event is that he makes zero A's in at least one semester. So, not $H =$ "he makes zero A's in at least one semester."

1.4   Some Probability Rules

In this section and also in Chapters 2 and 3, we will study some rules to simplify the determination of probabilities.

Probabilities satisfy the following rules:

Rule 1: $P(A) = 0$ if A is the impossible event $(A = \phi)$;

Rule 2: $P(A) = 1$ if A is the certain event $(A = S)$;

Rule 3: $P(A) = 1 - P(\text{not } A)$;

Rule 4: If A, B are mutually exclusive events,
$P(A \text{ or } B) = P(A) + P(B)$.
This is called the Addition Rule.

The addition rule can be extended to more than two events, so long as every pair of the events are disjoint (have no outcomes in common). That is, each pair of events is mutually exclusive.

Addition Rule: If $A_1, A_2, \ldots A_n$ are pairwise disjoint, then
$P(A_1 \text{ or } A_2 \text{ or } \ldots \text{ or } A_n) = P(A_1) + P(A_2) + \cdots + P(A_n)$.

Example 1.8. The following table gives the percentage of times each of the letters in the alphabet occurs in the English-language text:

A	7.3%	N	7.8%
B	0.9%	O	7.4%
C	3.0%	P	2.7%
D	4.4%	Q	0.3%
E	13.0%	R	7.7%
F	2.8%	S	6.3%
G	1.6%	T	9.3%
H	3.5%	U	2.7%
I	7.4%	V	1.3%
J	0.2%	W	1.6%
K	0.3%	X	0.5%
L	3.5%	Y	1.9%
M	2.5%	Z	0.1%

Choose one letter from English-language text.

Find:

(i) $P($vowel is chosen$)$

(ii) $P($consonant is chosen$)$

Solution:

(i) By the addition rule,

$$\begin{aligned}
P(\text{vowel is chosen}) &= P(A \text{ or } E \text{ or } I \text{ or } O \text{ or } U) \\
&= P(A) + P(E) + P(I) + P(O) + P(U) \\
&= .073 + .130 + .074 + .074 + .027 \\
&= .378.
\end{aligned}$$

(ii) We could use the addition rule to find $P($consonant is chosen$)$ but it is easier to use Rule 3.

$$\begin{aligned}
P(\text{consonant is chosen}) &= 1 - P(\text{vowel is chosen}) \\
&= 1 - .378 \\
&= .622.
\end{aligned}$$

Example 1.9. Our intuition tells us that if we were to look at the frequencies of leading digits for many types of data collections, then each of the digits between 1 and 9 inclusive would appear roughly an equal number of times. In fact, this turns out to be incorrect. Our intuition is wrong. It turns out that smaller digits are more likely than larger digits. This fact was first discovered when it was noticed that pages in books of mathematics tables tended to be dirtier and dog-eared closer to the front of the book. Such diverse data sets as census reports, accounting data, baseball statistics, and numbers appearing on newspaper pages have a non-zero first digit of 1 about 30% of the time in large samples. Numbers starting with 2 occur much more frequently than numbers starting with 6. There is a mathematical formula for the probabilities that appear here, but we will forego that formula in favor of giving the probability values computed from that formula. Details about the formula and the derivation are outside the scope of this book. First non-zero digits of many different kinds of data collections often conform to the following probabilities.

Non-Zero Leading Digit	Probability of that Digit
1	.301
2	.176
3	.125
4	.097
5	.079
6	.067
7	.058
8	.051
9	.046

Are we more likely to get a non-zero leading digit less than or equal to three or a non-zero leading digit greater than three?

Solution: By the addition rule,

$$P(\text{leading digit is} \leq 3) = .301 + .176 + .125 = .602.$$

By Rule 3,

$$P(\text{leading digit is} > 3) = 1 - P(\text{leading digit is} \leq 3) = 1 - .602 = .398.$$

Since $.602 > .398$, a non-zero leading digit less than or equal to three is more likely than a non-zero leading digit greater than three.

One application of the probability values given here is that they can assist in detecting data that have been fabricated such as numbers on an income tax form or accounting form. Fraudulent data tend to have far too few leading low digits and far too many leading high digits. When people fabricate data in financial documents, the proportions of non-zero leading digits from the fudged data deviate substantially from the actual probabilities of non-zero leading digits.

The next example is presented here because it requires that we focus on the details of a model.

Example 1.10. (A Model for Proposal Acceptance.) Suppose that in her lifetime, a woman will meet n potential husbands (proposers). After dating the first proposer, she must decide whether or not to accept him as a husband. If she accepts the first proposer, she never meets the other proposers. If she rejects him, he leaves her forever. She then meets and dates the second proposer. Either she accepts him as her husband or he leaves her forever. This process continues until she either picks a husband or runs out of proposers. If she had a crystal ball and could see all proposers in advance, she could rank them and select the best husband. But this is not the case because she meets these men one at a time. If she chooses a husband very early in the process, she is likely never to meet the best husband. If she waits too long, she might end up rejecting the best husband. What should she do? That is, what should her strategy be if she wants to *maximize her chances of selecting the best husband*? And what are the chances that if she uses that strategy she will get the best husband?

Suppose that as she receives each proposal she continually re-ranks the proposers. In reality, the number of potential husbands, n, is unknown in advance. But to help us understand the ranking idea, we will try the specific value of $n = 8$. We let 1 denote the best potential husband and 8 denote the worst. Suppose she had a crystal ball and could see the rankings after all 8 proposals and they were: 4 5 2 6 1 7 8 3. That is, the first proposer is ranked fourth overall, the second proposer is ranked fifth overall, etc.

After four proposers, even though their final rankings would be 4 5 2 6, based on the information she has to that point her ranking of those four would be 2 3 1 4. This is because the third man she meets is the best of the four, the first man she meets is the second best of the four, etc. Check your understanding of this idea by verifying that after five proposers, her rankings of these five would be 3 4 2 5 1.

It turns out that the form of the strategy that she should use to maximize her chances of selecting her best husband is the following:

> Reject the first k of the n proposers and then accept the first proposer after that who is better than all previous proposers (if there are any).

In the example, if $k = 1$, then the first proposer would be rejected. The first proposer after that who is better than all previous is the third proposer. He would be accepted and since he is ranked number two overall, he is therefore not the best.

If $k = 3$, then the first three proposers would be rejected. The first proposer after that who is better than all previous is the fifth proposer. In this case she gets the best husband.

If $k = 5$, then the first five proposers would be rejected. None of the proposers after that are better than all previous. So none of the proposers would be accepted and she doesn't get the best husband (she doesn't get a husband at all).

To determine the value of k in a few special cases, let's grind out some numerical answers. We begin by considering the situation where $n = 3$.

Suppose $k = 1$. So the first proposer is rejected and the first proposer after that who is better than all previous proposers is accepted. The equally probable sample space contains six possible outcomes:

$$
\begin{array}{c|cc}
1 & 2 & 3 \\
1 & 3 & 2 \\
2 & \mathbf{1} & 3 & x \\
2 & \mathbf{3} & 1 & x \\
3 & \mathbf{1} & 2 & x \\
3 & \mathbf{2} & 1 \\
\end{array}
$$

The proposer accepted is in bold type. We see that in three of the six equally likely outcomes which are marked with an x, she gets the best husband. Thus, P(she gets the best husband) $= 3/6 = .5$.

Suppose $k = 2$. So the first two proposers are rejected and the first proposer after that who is better than all previous proposers is accepted. Again the equally probable sample space contains six possible outcomes:

$$
\begin{array}{cc|c}
1 & 2 & 3 \\
1 & 3 & 2 \\
2 & 1 & 3 \\
2 & 3 & \mathbf{1} & x \\
3 & 1 & 2 \\
3 & 2 & \mathbf{1} & x \\
\end{array}
$$

The proposer accepted is in bold type. We see that in two of the six equally likely outcomes which are marked with an x, she gets the best husband. Thus, P(she gets the best husband) $= 2/6 = .33$.

Summary for $n = 3$:

k	P(she gets the best husband)
1	.5
2	.33

The chances of selecting the best husband are greatest when $k = 1$. So if $n = 3$, she should reject the first proposer and accept the next proposer after that who is better than all previous proposers. If she does that, the probability that she gets the best husband equals .5. We could also have included the cases where $k = 0$ and $k = 3$, but these would clearly not give the best strategy.

The value of k for which her chances of selecting the best husband are maximized depends on the value of n. You will be asked to find the value of k for which her chances of selecting the best husband are maximized when $n = 4$ in Problem 22 at the end of this chapter.

It turns out that when n is very large, her best strategy is to reject approximately the first 37% of the proposers and to accept the first proposer after that who is better than all previous proposers. Note that in Example 1.10, her best strategy is to reject the first $\frac{k}{n} = \frac{1}{3} = 33\%$ of the proposers.

Example 1.11. (Lottery as a Class Exercise.) The winning number in the lottery is a two-digit number from 00 to 99. To play the lottery each student in the class must write down a two-digit number between 00 and 99 inclusive which he believes will match the winning number. If a student matches the winning number, he will win a prize. Before revealing the winning number, your instructor will determine how frequently the digits from 0 to 9 were selected. Your instructor will call out the digits beginning with 0 and ending with 9 and each student will raise his hand(s) when his digits are called. So if the student selected 73 then the student raises one hand when 3 is called and one hand when 7 is called. If the student selected 33 then the student raises both hands when 3 is called. Then the instructor will count up the total number of hands that were raised for each digit. In addition to having a lottery, the class will have a frequency list for the selected digits from 0 to 9. Now your instructor will give you the winning number for the lottery.

1.5 Problem Solving

Problem solving provides students with a positive learning experience. The tools which are developed along the way can then be transferred to other areas of study and endeavor. Probability provides a wonderful vehicle for students to learn problem solving. It can be explained on a basic level without being too demanding technically. And applications of probability appear in most fields of study. Students are provided with an opportunity to develop their problem-solving skills while seeing how some real world situations are modeled.

While problem solving is rewarding and beneficial, it can at times be frustrating. What do we do when we reach a stumbling block? First of all, it is important to spend an adequate amount of time on the problem. Read the question carefully. It doesn't make much sense to spend two hours trying to solve a problem when you have taken ten seconds to misread the question. Make notes to glean the information from the problem as you read along. Let the problem sink in. Very often it will help to try several approaches. If you have been working on a problem for a long time without much success, put the problem aside and come back to it later. What frequently happens is that the solution will fall out on the next attempt. Even though you are not working directly on the problem between attempts, your subconscious is working on that problem. The more problem solving you do, the better your problem-solving skills will develop. It is important not to give up too soon and to maintain a positive attitude. You have to believe you are going to be successful. And the more frequently you are successful, the more positive your attitude will be when you approach the next problem. But even if you don't solve a problem, you learn by making a serious attempt because you are developing your problem-solving skills which will help you on future problems. You can sometimes learn more from a problem on which you make an unsuccessful attempt than from a problem that you are able to solve in a few seconds.

When you are attempting to solve a problem, its a good idea to delay looking at the answer until you are confident of your solution. Don't wait for your instructor to solve the problem in class. Understanding a solution that someone else has provided and coming up with the solution yourself are not the same. If you don't attempt the problem by yourself you are missing out on an important part of the problem-solving process.

I find it interesting that some students who are unable to solve a single homework problem in their dormitory room can somehow solve problems when they sit in my office even when I provide no information to them.

Some of the following techniques will be helpful in problem solving.

1. Act out the problem.

2. Relate a new problem to a familiar problem.

3. Look for a pattern.

4. Use a list, chart, or sketch.

5. Solve an equivalent but simpler problem.

Acting out a problem enables a student to become an active participant in problem solving. Problem 6 at the end of Chapter 1 is difficult for many students. This problem states, "Three letters written to individuals Smith, Jones, Clark are placed, at random, into envelopes addressed to Smith, Jones, Clark, with one letter in each envelope." The first part of the problem asks the student to "find the probability that none of the letters go into the correct envelope." When I discuss this problem in class, I take three envelopes labelled Smith, Jones, Clark and place them on the desk. Then I take three note cards labelled Smith, Jones, Clark, which represent the three letters and randomly distribute the cards in the envelopes by turning the cards over and shuffling them before distributing them. This gives the students one outcome in the sample space and completing the sample space then becomes much easier for them.

Another technique for problem solving is to relate the problem to a familiar one. Consider again Problem 6 at the end of Chapter 1. An examination of Example 1.3 reveals that if we replace dormitories with envelopes and signs with letters, we have Problem 6(i). By reading the examples in the book carefully, we develop a good foundation for categorizing problems. Problems don't usually match up quite as well as Example 1.3 and Problem 6(i), but there are often similar patterns that we should watch for.

Making a chart or table or sketch can be an effective tool in solving problems. Look at Problem 14 at the end of Chapter 1. "A single elimination tennis tournament has 4 players. Assume that the players can be ranked from 1 to 4, with player 1 always able to defeat player 2, player 2 always superior to player 3, etc. If the initial pairings for the tournament are assigned at random, what is the probability that the rank of the runner-up is 2?" So the players are paired up randomly at first and at each stage only the winner continues to compete. A diagram listing one possible outcome will be helpful:

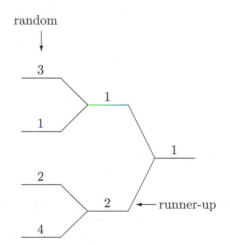

In this outcome the rank of the runner-up is 2. We would proceed by listing all outcomes in the sample space.

A famous dictum of Mathematician George Pólya's is that if you can't solve a problem, then there is an easier problem that you can solve—find it! Refer to Problem 35 in Chapter 2.

> A *palindrome* is a sequence of characters that reads the same forward or backward. For example, the words *madam* and *rotator* are palindromes. The number 7448447 is a 7-digit palindrome. What is the probability that a randomly selected 7-digit number is a palindrome?

At this point we would have to solve this problem by enumeration or some clever counting approach whereas the material in Chapter 2 will give us a more direct method for solution. It sometimes helps to look at a smaller version of the problem by cutting down on the number of possible outcomes. Even after we develop a solution to the original problem, having a solution to the smaller version allows us to check our answer. In the palindrome problem, we could ask the question: "What is the probability that a randomly selected 2-digit number is a palindrome? Since

$$S = \{00, 01, 02, 03, 04, 05, 06, 07, 08, 09\},$$

the answer is .1 since 00 is the only palindrome here. Next, we might ask the question, "What is the probability that a randomly selected 3-digit number is a palindrome?" Perhaps we might see a pattern by using this approach. If we don't observe a pattern but believe that we have found a valid solution when using a technique in Chapter 2, we can get a partial check on that solution by seeing if the application of our technique to the 2-digit and 3-digit numbers gives the correct answer.

For more information on problem solving, see reference [55] in the Bibliography.

1.6   Problems

1. In Example 1.2, if we were to repeat the experiment six times, would the event A necessarily occur exactly twice? What are the possible values for the number of times that A might occur?

2. Box I contains chips numbered 1, 2, 3 and box II contains chips numbered 1, 2, 3. One chip is selected at random from each box.

   (i) Is this the same as selecting two chips at random, with replacement, from box I?

   (ii) If $A =$ "average of the numbers drawn is 1.5," find $P(A)$.

3. Two William and Mary students arrive late for the math final exam and give the professor the phony excuse that their Mercedes Benz had a flat tire. "Well," the professor said, "each one of you write down on a piece of paper which of the four tires was flat." What is the probability that both students pick the same tire?

   (We will solve this problem using different methods in Chapters 2 and 3.)

4. At Brown University each student has to take two final exams. There are four possible times for finals. They are Monday a.m., Monday p.m., Tuesday a.m., Tuesday p.m. What is the probability that a student with a randomly selected schedule has her exams in consecutive exam periods? (Monday p.m. and Tuesday a.m. are considered to be consecutive exam periods.)

5. Excerpt from the TV show, *The Rockford Files*:

   ROCKFORD: There are only two doctors in town. The chances of both autopsies being performed by the same doctor is 1/2.

   REPORTER: No, that is only for one autopsy. For two autopsies, the chances are 1/4.

   ROCKFORD: You're right.

   Was Rockford right to agree with the reporter? Explain why or why not. Assume both doctors are equally likely to perform an autopsy.

6. Three letters written to individuals Smith, Jones, Clark are placed, at random, into envelopes addressed to Smith, Jones, Clark, with one letter in each envelope.

   Find the probability that

   (i) None of the letters go into the correct envelope;

   (ii) Exactly two letters go into the correct envelopes.

7. Tony Gwynn has a lifetime batting average of .33. That is, he averages one hit in three official at bats in the long run. If he has two official at bats in a rain-shortened game, find the probability that he gets at least one hit in this game.

8. Three balanced dice are rolled. Estimate the probability of getting three different faces by performing 10 simulations using the random number target.

9. Describe how to estimate $P(A)$ for Example 1.1 without ignoring any numbers from the random number target.

10. A baseball player batted twice in a game. For each of the following pairs of events, state whether or not the two events are mutually exclusive. Then state whether of not the two events are complementary.

   (i) $A =$ "hit on first at bat"; $B =$ "hit on second at bat";
   (ii) $A =$ "exactly one hit"; $B =$ "exactly two hits";
   (iii) $A =$ "at least one hit"; $B =$ "zero hits."

11. Two chips are selected at random without replacement from a bowl containing four chips numbered 1, 2, 3, 4. What is the probability that the average of the two numbers selected:

   (i) Equals 2?
   (ii) Exceeds 2?
   (iii) Is greater than or equal to 2? Use $P(A \text{ or } B) = P(A) + P(B)$.
   (iv) Is less than 2? Use $P(C) = 1 - P(\text{not } C)$.

12. Four William and Mary students are waiting in line in front of a ticket window to buy $50 tickets to a "Pearl Jam" concert. Two of the students have $50 bills and the other two have $100 bills. Initially, the cashier has no change. What is the probability that those with $50 bills will be positioned in the line in such a way that the cashier always has enough change so that the four students will be able to buy their tickets? Assume the students are positioned in line at random. Can you relate the sample space in this problem to the sample space in an example in this chapter? Which example?

13. The famous nineteenth-century American astronomer, Simon New-comb, amused himself by playing a game of solitaire related to the following question. A deck of cards contains 3 cards, numbered 1, 2, 3. The deck of cards is first shuffled. Then the 3 cards are drawn one at a time and are put into a pile as long as the number on the card drawn is lower than the number on its immediate predecessor; a new pile is always started when a higher card than its immediate predecessor is drawn. Find the probability that there are two piles.

14. A single elimination tennis tournament has four players. Assume that the players can be ranked from 1 to 4, with player 1 always able to defeat player 2, player 2 always superior to player 3, etc. If the initial pairings for the tournament are assigned at random, what is the probability that the rank of the runner-up is 2?

15. Draw two chips, at random, with replacement from a box containing two chips numbered 1 and 2. Let $A = $ "sum of numbers exceeds 2."

    (i) Find $P(A)$.

    (ii) Why must you count ordered pairs? What happens when you count unordered pairs? Try it.

16. For four years, two college teams play each other to decide the state championship. The winner gets possession of a silver trophy. Suppose that during the four years, Richmond has won twice and Tech has won twice. What is the probability that the trophy has changed hands more than once?

17. The new freshman dormitory at Christopher Newport University has all double rooms. All but two of these rooms were occupied by students who indicated specific roommate preferences on their applications. There were two white students and two black students who did not indicate roommate preference on their applications. They were paired up for the remaining two rooms by the dormitory manager. Neither of these rooms turned out to be of mixed race. If two white students and two black students are assigned at random to two double rooms, what is the probability of getting zero mixed-race rooms?

18. Suppose Albert Belle bats four times each game in a series of three games. Describe the complementary event to each of the following events:

(i) Albert Belle gets 4 hits in every game in the series;

(ii) Albert Belle gets at least 1 hit in every game in the series;

(iii) Albert Belle gets at least 1 hit in the series;

(iv) Albert Belle gets 4 hits in at least 1 game in the series.

19. Suppose there are 65 students in class the day of the lottery (see Example 1.11). A student in the class wants to determine the probability that the professor won't have to give away any prizes. He reasons as follows:

$$P(\text{at least one student matches the winning number})$$
$$= P(\text{student 1 matches or student 2 matches or} \ldots$$
$$\text{or student 65 matches})$$
$$= P(\text{student 1 matches}) + P(\text{student 2 matches})$$
$$+ \cdots + P(\text{student 65 matches})$$
$$= \frac{1}{100} + \frac{1}{100} + \cdots + \frac{1}{100} = \frac{65}{100}.$$

Thus, $P(\text{no one matches the winning number}) = 1 - P(\text{at least one student matches the winning number}) = 1 - \frac{65}{100} = .35$.

Is this solution correct? Why or why not?

20. When a bullet is fired it becomes scored with minute grooves produced by imperfections in the gun barrel. Appearing as parallel lines, these grooves can provide evidence for matching a bullet with a gun, since repeated firing of the same gun will produce bullets having a similar pattern of grooves. Suppose a bullet is recovered from a crime scene along with a suspect's gun. Under a microscope, a grid with 3 cells, numbered 1, 2, 3, is superimposed over the bullet. Then the suspect's gun is fired providing a test bullet to be compared with the evidence bullet. Suppose the evidence bullet has two cells which have grooves, and the test bullet has two cells which have grooves. What is the probability that exactly one of these grooved-cell locations matches? Does the answer to this question depend on which two cells in the evidence bullet have grooves? (In an actual case, there would be more than 3 cells per bullet).

21. A caller to a sports talk show on radio station KDKA in Pittsburgh commented that the Pittsburgh Steeler Football Team had been guilty of poor tackling. He blamed this on the team coaching staff.

The host responded by saying that you could have good coaching yet poor tackling. "In other words," he said, "coaching and tackling are mutually exclusive." Which one of the following statements describes what the host should have said?

(a) Good coaching and poor tackling are mutually exclusive.

(b) Coaching and tackling are not mutually exclusive.

(c) Good coaching and poor tackling are not mutually exclusive.

22. Refer to Example 1.10. Find the value of k which gives the maximum probability of selecting the best husband when $n = 4$. Show that the probability of selecting the best husband for this value of k is .458.

23. A teacher has a small class with three students. There are three seats in the classroom numbered 1, 2, 3. Although the teacher has prepared a seating chart, the students have already seated themselves randomly. The teacher calls off the name of the student who belongs in seat 1. This student vacates the seat he or she is currently occupying and takes his or her rightful seat. If this displaces a student already in the seat, that student stands in front of the room until he or she is assigned a seat. The teacher does this for each seat in turn. After two names have been called, what is the probability that there is one student standing at the front of the room?

24. In 1983, the best college player available in the National Basketball Association draft was Ralph Sampson from the University of Virginia. Two teams, Indiana and Houston, were tied for having the worst records the previous season and were to get an equal chance at getting the first pick in the draft. A coin tossing ceremony was held in the office of league commissioner Larry O'Brien. It consisted of two flips of a silver dollar. The purpose of the first flip was to determine who would call the second flip. The purpose of the second flip was to determine who would get the first pick in the draft and therefore the rights to negotiate with Ralph Sampson. The owner of the Houston Rockets, Charlie Thomas, won the first toss and then correctly called "heads" on the second toss. Was the first toss superfluous? Why?

2

Conditional Probability and the Multiplication Rule

2.1 Conditional Probability

Sometimes we are concerned with probabilities about some portion of the sample space rather than the entire sample space. Here are two examples. The probability that a person has an annual income over $100,000 would be different than the probability that a college graduate has an annual income over $100,000. As of June 27, 2000, John Vander Wal had a season batting average of .292, but his batting average against left-handed pitchers was .188. In both of these examples, we are reducing the sample space. For the first example, the reduced sample space consists of college graduates and in the second example the reduced sample space consists of left-handed pitching opposition. The following example illustrates the concept of conditional probability.

Example 2.1. One chip is selected at random from a box containing five chips numbered 1, 2, 3, 4, 5. So $S = \{1, 2, 3, 4, 5\}$ is an equally probable sample space. What is the probability of selecting a 1? If we let $B = \{1\}$, then $P(B) = \frac{1}{5}$. Now suppose we are told that the outcome is an odd number. Let $A = \{1, 3, 5\}$. We are given that the outcome is in A. Now what is the probability of getting a 1? The answer is $\frac{1}{3}$. We write this as $P(B|A) = \frac{1}{3}$. The vertical line separating events B and A means "given" and $P(B|A)$ is called the conditional probability of B given A.

We reduce the sample space from S to A while maintaining the same relative probabilities between outcomes. Since each outcome in S has an equal probability, the same must be true in the reduced sample space A. That is, we increase the probability of each outcome in A proportionately. So each outcome 1, 3, 5 has probability $\frac{1}{5}$ relative to S but is increased to $\frac{1}{3}$ relative to A. Had one outcome in S been twice as probable as another outcome, these same relative probabilities would carry over to the reduced sample space.

Another method for finding a conditional probability is by using the formula:

$$P(B|A) = \frac{P(B \text{ and } A)}{P(A)},$$

where $A \neq \phi$.

The probabilities on the right-hand side of the equation are relative to S. To solve Example 2.1 using this formula, we first must find the event "B and A." But the event "B and A" $= \{1\}$ since "B and A" includes outcomes common to both event B and event A. Relative to the original sample space S,

$$P(B \text{ and } A) = P(\{1\}) = \tfrac{1}{5}, \text{ and}$$

$$P(A) = P(\{1,3,5\}) = \tfrac{3}{5}.$$

The formula gives

$$P(B|A) = \frac{P(B \text{ and } A)}{P(A)} = \frac{1/5}{3/5} = \frac{1}{3},$$

which agrees with our answer arrived at by reducing the sample space to A.

We can find $P(B|A)$ using either method but unless the problem is very simple with an equally probable sample space, using the formula is preferable to reducing the sample space.

Example 2.2. One chip is selected from a box containing three chips numbered 1, 2, 3 where the 3 is twice as likely to be selected as either the 1 or the 2. If $B =$ "outcome greater than 1," and $A =$ "odd outcome," find $P(B|A)$.

Solution: To get an equally probable sample space, we need to list 3 twice. To ensure accurate bookkeeping we use $3a, 3b$. So $S = \{1, 2, 3a, 3b\}$ is an equally probable sample space.

The reduced sample space is $A =$ "odd outcome" $= \{1, 3a, 3b\}$. Note that 3 is still twice as likely as 1 just as it was in S. So $P(B|A) = \frac{2}{3}$ since two of the elements in A are greater than 1.

To solve this example using the formula method rather than the reduced sample space, we look at the probabilities relative to the original sample space S.

$$B = \{2, 3a, 3b\};$$
$$A = \{1, 3a, 3b\};$$
$$B \text{ and } A = \{3a, 3b\};$$
$$P(B|A) = \frac{P(B \text{ and } A)}{P(A)} = \frac{2/4}{3/4} = \frac{2}{3}.$$

2.2 Multiplication Rule

Multiplying both sides of the equation $P(B|A) = \frac{P(B \text{ and } A)}{P(A)}$ by $P(A)$ gives

$$P(B \text{ and } A) = P(A)P(B|A).$$

This is called the multiplication rule, and it gives us a method for finding $P(B \text{ and } A)$ when the conditional probability is known.

The multiplication rule also holds when we have more than two events. For three events, the multiplication rule is

$$P(A \text{ and } B \text{ and } C) = P(A)P(B|A)P(C|A \text{ and } B).$$

In the next few examples we will rephrase an event in terms of its component events which are connected by the word "and." The multiplication rule can then be applied.

Example 2.3. The winning number in a class lottery is a two-digit, computer-generated number from 00 to 99. To play the lottery, each student in the class writes down a two-digit number between 00 and 99 inclusive. What is the probability that none of the students in the class will match the winning number if there are 65 students in the class?

Solution: Let

$A =$ 1st student does not match the winning number.
$B =$ 2nd student does not match the winning number.
$C =$ 3rd student does not match the winning number, etc.

Then, A and B and C and $\ldots$ = no student wins the lottery.

$$
\begin{aligned}
P(\text{no student wins the lottery}) &= P(A \text{ and } B \text{ and } C \text{ and} \ldots) \\
&= P(A)P(B|A)P(C|A \text{ and } B)\cdots \\
&= \frac{99}{100} \cdot \frac{99}{100} \cdot \frac{99}{100} \cdots \\
&= \left(\frac{99}{100}\right)^{65} \\
&= .520 \text{ since there are 65 students.}
\end{aligned}
$$

So there is slightly better than a 50–50 chance that no one wins the lottery.

Example 2.4. Urns I, II, III each contain four chips numbered 1, 2, 3, 4. One chip is selected at random from each urn. Find the probability of getting three different numbers.

It would be possible but somewhat tedious to solve this problem by enumerating outcomes. For example, $S = \{(1,1,1),(1,1,2),(1,1,3)\ldots (4,4,4)\}$ where the first entry in each ordered triple corresponds to the chip number from I, the second entry in each ordered triple corresponds to the chip number from II, and the third entry in each ordered triple corresponds to the chip number from III.

It is much easier to solve this problem by using the multiplication rule than by direct enumeration.

Solution: Let

$$
\begin{aligned}
A &= \text{``all 3 numbers drawn are different''}; \\
B &= \text{``any number is drawn from I''}; \\
C &= \text{``number from II is different than number from I''}; \\
D &= \text{``number from III is different than numbers from I and II.''}
\end{aligned}
$$

Then

$$
\begin{aligned}
A &= B \text{ and } C \text{ and } D; \\
P(A) &= P(B \text{ and } C \text{ and } D); \\
P(A) &= P(B)P(C|B)P(D|B \text{ and } C); \\
P(A) &= \frac{4}{4} \cdot \frac{3}{4} \cdot \frac{2}{4} = .375.
\end{aligned}
$$

Example 2.5. (Birthday Problem.)   Suppose there are 65 students in a room (no twins). What is the probability that all 65 have different birthdays? Assume all 365 birthdays are equally likely. Imagine 65 boxes, one for each student. Each box contains 365 chips which correspond to the 365 days of the year. We think in terms of selecting one chip from each box which corresponds to selecting a birthday, at random, for each student.

Solution: Let

$$
\begin{aligned}
A &= \text{``all 65 students have different birthdays''};\\
B &= \text{``any birthday is selected for student 1''};\\
C &= \text{``birthday for student 2 is different than}\\
  &\quad \text{birthday for student 1''};\\
D &= \text{``birthday for student 3 is different than}\\
  &\quad \text{birthdays for students 1 and 2,'' etc.}
\end{aligned}
$$

Then

$$
\begin{aligned}
A &= B \text{ and } C \text{ and } D \text{ and } \ldots;\\
P(A) &= P(B \text{ and } C \text{ and } D \text{ and } \ldots);\\
P(A) &= P(B)P(C|B)P(D|B \text{ and } C)\ldots \text{ by the multiplication rule.}\\
P(A) &= \frac{365}{365}\cdot\frac{364}{365}\cdot\frac{363}{365}\cdots\frac{301}{365} = .01 \text{ by computer.}
\end{aligned}
$$

So,

"Not A" = "At least two students in the room have the same birthday."

$$P(\text{Not } A) = 1 - P(A) = 1 - .01 = .99.$$

So it is almost certain that at least two students in the room will have the same birthday. Using the same type of reasoning it is possible to show that we need only 23 students in a room to have better than a 50-50 chance that at least two students have the same birthday. Like many numerical results in probability the answer is counter-intuitive.

Example 2.6. Johnny Carson, on hearing about the birthday problem, once observed during the Tonight Show that there were about 120 people in his audience. He asked all audience members who shared his birthday of October 23 to raise their hands. To Johnny's surprise, there were no raised hands. Johnny's mistake was that while the probability of at least two audience members having the same birthday is large, the probability that at least one audience member has a particular birthday which matches his is quite small.

To see this, let

$$A = \text{"No audience member has an October 23 birthday,"}$$

$$B = \text{"first audience member does not have an}$$
$$\text{October 23 birthday,"}$$

$$C = \text{"second audience member does not have an}$$
$$\text{October 23 birthday,"}$$

$$D = \text{"third audience member does not have an}$$
$$\text{October 23 birthday," etc.}$$

So,

$$A = B \text{ and } C \text{ and } D \text{ and } \ldots;$$
$$P(A) = P(B \text{ and } C \text{ and } D \text{ and } \ldots)$$
$$= P(B)P(C|B)P(D|B \text{ and } C)\ldots$$
$$= \frac{364}{365} \cdot \frac{364}{365} \cdot \frac{364}{365} \cdots = \left(\frac{364}{365}\right)^{120} = .72.$$

$$\text{"Not } A\text{"} = \text{"At least one audience member has a}$$
$$\text{birthday on October 23."}$$

$$P(\text{Not } A) = 1 - P(A) = 1 - .72 = .28.$$

Thus, the chances of Johnny finding a studio audience member with the same birthday as his is only .28.

The following example shows how to use probability in modeling an actual situation.

Example 2.7. During the 120-day period between November 1968 and February 1969, there were 22 commercial airliners hijacked to Cuba. On a day when there were two hijackings, the *New York Times* regarded the occurrence of more that one hijacking on the same day as a "sensational and improbable coincidence."

Model: Suppose 22 balls are tossed into 120 boxes at random. Label the 120 boxes: November 1, 1968, November 2, 1968, ..., February 28, 1969. When a ball lands in a box it corresponds to a hijacking on that day.

$$P(\text{22 balls fall into different boxes}) =$$
$$P(\text{never more than one hijacking on a day}) = \frac{120}{120} \cdot \frac{119}{120} \cdots \frac{99}{120}$$
$$= .1285.$$

So,
$$P(\text{some box contains more than 1 ball}) =$$
$$P(\text{two or more hijackings on the same day}) = 1 - .1285$$
$$= .8715.$$

We see that the occurrence of more than one hijacking on the same day is not improbable.

Example 2.8. An advertisement in *The New Yorker* of April 22, 1967 began:

WILL THE REAL MILLER HIGH LIFE DRAFT BEER
PLEASE STEP FORWARD

"We've tried our famous Miller High Life taste test on many a 'beer expert'; pouring three glasses, one from the tap (T), one from the can (C), and one from the bottle (B). And then we've asked which is which. Result? No one, up to now, has identified the three correctly. Why? *All three glasses* have the same distinctive Miller High Life flavor."

Comment on the ad: If the three ways of packaging the beer lead to indistinguishable glasses, then the probability that one beer expert does not identify the three correctly is $\frac{5}{6}$. This is similar to Example 1.3. We list the equally likely outcomes in the sample space.

Glass 1	Glass 2	Glass 3
T	C	B
T	B	C
B	C	T
B	T	C
C	T	B
C	B	T

Only one of the six outcomes corresponds to identifying the three glasses correctly.

The ad claims that no one, up to now, has identified the three correctly. The ad doesn't specify how many experts participated. By the multiplication rule, if there were five experts,

$$P(\text{no one identified the 3 correctly}) = \left(\frac{5}{6}\right)^5 = .40,$$

while if there were 20 experts,

$$P(\text{no one identified the 3 correctly}) = \left(\frac{5}{6}\right)^{20} = .026.$$

So if there were 20 experts it would be very unlikely to get the observed outcome. It is likely that not many experts participated.

2.3   Independence

The events A, B are said to be independent events if the occurrence or non-occurrence of event A does not affect the probability of occurrence of event B. That is, $P(B|A) = P(B| \text{ not } A) = P(B)$.

The multiplication rule states:

$$P(A \text{ and } B) = P(A)P(B|A).$$

So when A, B are independent events, the multiplication rule becomes

$$P(A \text{ and } B) = P(A)P(B).$$

Either of two equations can be used to check for independence:

(i) $P(B|A) = P(B)$ (equivalently $P(A|B) = P(A)$);

(ii) $P(A \text{ and } B) = P(A)P(B)$.

Equation (ii) is usually easier to check than (i).

Note: Don't confuse independent events with mutually exclusive events. Problems 17 and 20 at the end of this chapter as well as the following example should be helpful in keeping these concepts separate.

Example 2.9. A balanced die is rolled. Let $A =$ "an odd number of dots appear," $B =$ "number of dots exceeds 3," $C =$ "number of dots exceeds 4."

(i) Are A, B independent events?

$A = \{ 1, 3, 5 \}$, $B = \{ 4, 5, 6 \}$, A and $B = \{ 5 \}$;

$P(A) = \dfrac{3}{6}$, $P(B) = \dfrac{3}{6}$, $P(A \text{ and } B) = \dfrac{1}{6}$.

Check condition (ii) above.

Does $P(A \text{ and } B) = P(A)P(B)$?

Does $\dfrac{1}{6} = \dfrac{3}{6} \cdot \dfrac{3}{6}$?

The answer is no, so that A, B are not independent events.

(ii) Are A, C independent events?

$$A = \{1, 3, 5\}, \quad C = \{5, 6\}, \quad A \text{ and } C = \{5\};$$

$$P(A) = \frac{3}{6}, \qquad P(C) = \frac{2}{6}, \quad P(A \text{ and } C) = \frac{1}{6}.$$

Does $P(A \text{ and } C) = P(A)P(C)$?

Does $\dfrac{1}{6} = \dfrac{3}{6} \cdot \dfrac{2}{6}$?

The answer is yes, so that A, C are independent events. Note that A, C are not mutually exclusive because both events have the outcome 5 in common.

Example 2.9 shows that whether or not events are independent is not always obvious. An example where the question of independence is obvious follows. Two chips are drawn from a box containing two red and two blue chips and A denotes the event "a red chip is selected on the first draw," and B denotes the event "a red chip is selected on the second draw." If the chips are drawn with replacement then A, B are independent events because the outcome of the first draw does not affect the probability of getting a red chip on the second draw. However, if the drawings are without replacement, then A, B are not independent events because the outcome on the first draw would affect the probability of getting a red chip on the second draw.

Outcomes on successive coin tosses provide another example of independence. Suppose a fair coin is tossed ten times. Let $A =$ "heads on the first nine tosses" and $B =$ "heads on the tenth toss." Then $P(B|A) = P(B)$ and A, B are independent events. Getting nine heads in a row does not mean that the chances of a tail on the tenth toss increases. The coin has no memory.

Sometimes we just assume independence based on the statement of the problem. For example, if teams A, B play two games, we might assume that the events "team A wins the first game" and "team A wins the second game" are independent events. There is no reason to believe that the outcome of the first game would affect the probability of team A winning the second game.

The multiplication rule for independent events can be extended to the case where there are more than two events. For example, if A, B, C are independent events, then

$$P(A \text{ and } B \text{ and } C) = P(A)P(B)P(C).$$

Example 2.10. (Class Exercise.) Do you expect males or females to have more sisters? A classroom experiment can be instructive here. Suppose each student in the class writes on a notecard his or her sex and the number of sisters he or she has. Then we will compare the responses from those cards written by male students with those written by female students. We might even calculate the average number of sisters for the male and female students. To avoid confusion, define a sister as a woman who has the same two biological parents as the respondent. If the class is large enough, the data we collect should provide a good indication of the answer to our question.

Example 2.11. It has been said that a monkey sitting at a keyboard will eventually type out the entire works of Shakespeare if he continually hits keys at random. This statement is a story version of a famous law in probability theory. The comedian Bob Newhart on his record album "The Button-Down Mind Strikes Back!" reflects on a room full of these monkeys and how a staff of human inspectors would be needed to check on their progress. Let's see if we can understand how to simulate a Shakespeare monkey.

Table 1 lists the relative frequencies of letters, apostrophe, and space taken from Act III of Hamlet.

We see that the letter M appears about 2.5% of the time. We take these long-run relative frequencies to be our probabilities. By generating characters according to these probabilities we could produce the writings of a "first-level Shakespeare monkey." We would carry out a simulation on a computer where in the long run, an A would appear about 5.8% of the time, a B about 1.2% of the time, etc. The problem here is that while the overall percentages of characters would be consistent with those in Act III of Hamlet, the letters are unrelated in sequence. For example, if a Q is selected as the first character, using this simulation technique, the chance that the next letter is U is .029 whereas it should be 1 since U

Character	Relative Frequencies	Character	Relative Frequencies
A	.058	O	.073
B	.012	P	.012
C	.017	Q	.001
D	.031	R	.045
E	.093	S	.053
F	.018	T	.072
G	.014	U	.029
H	.050	V	.009
I	.049	W	.020
J	.001	X	.001
K	.007	Y	.022
L	.035	Z	.001
M	.025	'	.006
N	.049	space	.197

Total
1.000

Table 2.1. Relative frequencies of letters, apostrophe, and space taken from Act III of Hamlet.

always follows Q in the English language. In other words, the selection of letters by the "first-level Shakespeare monkey" is independent.

What we need for a "second-level Shakespeare monkey" are conditional probabilities. That is, if a W is selected as the first character, what are the conditional probabilities of getting A, B, C, etc. for the next character? Since probabilities are long-run proportions,

$$
\begin{aligned}
P(A|W) &= \text{long-run proportion of times that } A \text{ follows } W \\
&= (\text{number of } WA\text{'s})/(\text{number of } W\text{'s}).
\end{aligned}
$$

We know that $P(U|Q) = 1$ since U must follow Q and $P(D|S) = 0$ since D never follows S in the English language. The simulation carried out by the second-level Shakespeare monkey consists of first selecting a letter character using the probabilities of letters from above and thereafter using conditional probabilities of the form $P(y|x)$. If for example $P(D|A) = .1$, then whenever an A appears, the simulation would select D for the next letter 10% of the time.

The same idea can be extended to higher levels. For example,

$$P(A|QU) = \text{long-run proportion of times that } A \text{ follows } QU$$
$$= (\text{number of } QUA\text{'s})/(\text{number of } QU\text{'s}).$$

We know that $P(A|QU) > 0$ while $P(G|QU) = 0$ because A can follow QU but G never follows QU. The simulation carried out by the third-level Shakespeare monkey consists of first selecting a letter character using the relative frequencies of single characters and then using a conditional probability of the form $P(y|x)$ to generate the second letter followed thereafter by conditional probabilities of the form $P(z|xy)$.

Suppose we want to find the probability of forming the word *ATE*. The first-level monkey would use the formula:

$$P(ATE) = P(A)P(T)P(E)$$

because he treats the selection of characters as independent events. The second-level monkey would use the formula:

$$P(ATE) = P(A)P(T|A)P(E|T);$$

in other words he uses only the previous character in forming conditional probabilities after the first letter is chosen. The third-level monkey would use the formula:

$$P(ATE) = P(A)P(T|A)P(E|AT);$$

in other words he uses two previous characters in determining conditional probabilities after the first two characters are chosen.

The following are computer generated samples from the first-, second-, third-, and fourth-level Shakespeare monkeys. The first-level monkey does not get a single correctly spelled word. The fourth-level monkey shows substantial improvement; the message is still garbled but most of the letter combinations are words. Higher level monkeys would no doubt be able to generate something intelligible.

First-Level Shakespeare Monkey

AAOAAORH ONNNDGELC TEFSISO VTALIDMA POESDHEMHIESWON
PJTOMJ ITL FIM TAOFERLMT O NORDEERH HMFIOMRETW OVRCA
OSRIE IEOBOTOGIM NUDSEEWU WHHS AWUA HIDNEVE NL SELTS
ILY L CPATSDM EOLF OSTOM

Second-Level Shakespeare Monkey

AROABLON MERMAMBECRYONSOUR T T ANED AVECE AMEREND TIN NF MEP HIN FOR'T SESILORK TITIPOFELON HELIORSHIT MY ACT MOUND HARCISTHER K BOMAT Y HE VE SA FLD D E LI Y ER PU HE YS ARATUFO

Third-Level Shakespeare Monkey

TO HOIDER THUS NOW GOONS ONES NO ITS WHIS KNOTHIMEN AS TOISE MOSEN TO ALL YOURS YOU HOM TO TO LON ESELICES HALL IT BLED SPEAL YOU WOUNG YEAT BE ADAMED MY WOME COUR TO MUSIN SWE PLAND NAVE PRES LAIN IFY YOUGHTS THAVE OF

Fourth-Level Shakespeare Monkey

A GO THIS BABE AND JUDGEMENT OF TIMEDIOUS RETCH AND NOT LORD WHAL IF THE EASELVES AND DO AND MAKE AND BASE GATHEM I AY BEATELLOUS WE PLAY MEANS HOLY FOOL MOUR WORK FROM INMOST BED BE CONFOULD HAVE MANY JUDGMENT

In 1935, Sir Arthur Eddington considered the idea of musical composition using this type of technique. In *New Pathways in Science*, he wrote the following rhyme:

> *There once was a brainy baboon*
> *Who always breathed down a bassoon*
> *For he said, "It appears*
> *That in billions of years*
> *I shall certainly hit on a tune"*

The next section depends on the material in Chapter 1 which deals with complementary events.

2.4 A Technique for Finding $P(A$ or B or C or ...)

We know from Chapter 1 that if events $A, B, C, \ldots$ are pairwise mutually exclusive then $P(A$ or B or C or $\ldots) = P(A) + P(B) + P(C) + \ldots$ by the addition rule. Suppose the events in question are not pairwise mutually exclusive. Then

$$
\begin{aligned}
P(A \text{ or } B \text{ or } C \text{ or } \ldots) \\
&= P(\text{at least one of } A, B, C \ldots \text{ occur}) \\
&= 1 - P(\text{none of the events } A, B, C \ldots \text{ occur}) \\
&= 1 - P(\text{"not } A\text{" and "not } B\text{" and "not } C\text{" and} \ldots).
\end{aligned}
$$

At this point, the multiplication rule can be applied. If, in addition, the events connected by the word "and" are independent, we can write:

$$
\begin{aligned}
P(A \text{ or } B \text{ or } C \text{ or} \ldots) \\
&= 1 - P(\text{"not } A\text{" and "not } B\text{" and "not } C\text{" and} \ldots) \\
&= 1 - P(\text{not } A)P(\text{not } B)P(\text{not } C) \cdots \\
&= 1 - [1 - P(A)][1 - P(B)][1 - P(C)] \cdots.
\end{aligned}
$$

This technique will be used in the next three examples.

Example 2.12. A fair coin is tossed three times. Find P(at least one head).

Solution:

$$
S = \{HHH, HHT, HTH, THH, HTT, THT, TTH, TTT\}.
$$

This is an equally probable sample space. It is easy to see by enumerating the outcomes in S which correspond to the event "at least one head" that P(at least one head) $= \frac{7}{8}$. But we want to examine this problem in greater depth.

We decompose the event "at least one head" in two different ways:

(i) Let A = "exactly one head," B = "exactly two heads," C = "exactly three heads."

(ii) Let H_1 = "head on 1st toss," H_2 = "head on 2nd toss," H_3 = "head on 3rd toss."

Observe that in (i), "at least one head" = A or B or C. Also observe that in (ii), "at least one head" = H_1 or H_2 or H_3. Make sure that the last two statements are clear to you.

Let's focus first on the solution to the problem using (i):

$$A = \{HTT, THT, TTH\}, \quad B = \{HHT, HTH, THH\}, \quad C = \{HHH\}.$$

In (i), the events A, B, C are pairwise mutually exclusive. That is, none of the pairs of events, A, B, or B, C, or A, C have any outcomes in common. Therefore, the addition rule applies and

$$P(\text{at least one head}) = P(A \text{ or } B \text{ or } C) = P(A) + P(B) + P(C).$$

This can be verified numerically by observing that

$$P(\text{at least one head}) = P(A \text{ or } B \text{ or } C) = \frac{7}{8},$$

while

$$P(A) + P(B) + P(C) = \frac{3}{8} + \frac{3}{8} + \frac{1}{8} = \frac{7}{8}.$$

We focus next on the solution to the problem using (ii):

$$
\begin{aligned}
H_1 &= \{HHH, HHT, HTH, HTT\}; \\
H_2 &= \{HHH, HHT, THH, THT\}; \\
H_3 &= \{HHH, HTH, THH, TTH\}.
\end{aligned}
$$

Observe that H_1, H_2, H_3 are not pairwise mutually exclusive so the addition rule does not apply here. So

$$P(H_1 \text{ or } H_2 \text{ or } H_3) \neq P(H_1) + P(H_2) + P(H_3).$$

This can be verified numerically by observing that

$$P(\text{at least one head}) = P(H_1 \text{ or } H_2 \text{ or } H_2) = \frac{7}{8},$$

while

$$P(H_1) + P(H_2) + P(H_3) = \frac{1}{2} + \frac{1}{2} + \frac{1}{2} = \frac{3}{2}.$$

Since the addition rule doesn't apply here, to solve P(at least one head) using (ii),

$$
\begin{aligned}
P(\text{at least one head}) \;&=\; P(H_1 \text{ or } H_2 \text{ or } H_3) \\[4pt]
&=\; 1 - P(\text{not } H_1 \text{ and not } H_2 \text{ and not } H_3) \\[4pt]
&=\; 1 - P(\text{not } H_1)P(\text{not } H_2)P(\text{not } H_3) \\[4pt]
&=\; 1 - \{[1 - P(H_1)][1 - P(H_2)][1 - P(H_3)]\} \\[4pt]
&=\; 1 - \left\{ \left[1 - \frac{1}{2}\right]\left[1 - \frac{1}{2}\right]\left[1 - \frac{1}{2}\right] \right\} \\[4pt]
&=\; 1 - \left(\frac{1}{2}\right)^3 \\[4pt]
&=\; \frac{7}{8}.
\end{aligned}
$$

So P(at least one head in 3 tosses of a fair coin) can be solved by enumeration, or by defining A, B, C as in (i), and then using the addition rule, or by defining H_1, H_2, H_3 as in (ii) and appealing to the rule for complementary events and then the multiplication rule. Suppose that we change the problem from P(at least one head in 3 tosses of a fair coin) to P(at least one head in 20 tosses of a fair coin). The only feasible way to solve this problem is to use the approach in (ii):

$$
\begin{aligned}
P(\text{at least one head in 20 tosses of a fair coin}) \\
= 1 - P(\text{no heads in 20 tosses of a fair coin}) \\
= 1 - P(\text{20 consecutive tails}) \\
= 1 - \left(\frac{1}{2}\right)^{20} \\
= .999999.
\end{aligned}
$$

Example 2.13. In 1978 Pete Rose set a National League record by hitting safely in each of 44 consecutive games. His lifetime batting average was .303 (slightly more than 30% of the time he got a hit). Also, assume he came to bat four times each game and his chances of getting a hit on each at bat did not depend on previous at bats. Find the probability that:

(i) He got at least one hit in a given game.

(ii) He got at least one hit in each of 44 consecutive games.

Solution:

(i) $P(\text{at least one hit in a game})$
$$= P(\text{``hit on } 1^{st} \text{ at bat'' or} \ldots \text{or ``hit on } 4^{th} \text{ at bat''}).$$

The addition theorem does not apply since the events "hit on 1st at bat" and "hit on 2nd at bat" are not mutually exclusive.

So,

$P(\text{at least one hit in a game})$

$= \quad 1 - P(\text{no hits in a game})$

$= \quad 1 - P(\text{``out on } 1^{st} \text{ at bat'' and} \ldots \text{and ``out on } 4^{th} \text{ at bat''})$

$= \quad 1 - P(\text{out on } 1^{st} \text{ at bat}) \cdots P(\text{out on } 4^{th} \text{ at bat})$

by the independence assumption

$= \quad 1 - [1 - P(\text{hit on } 1^{st} \text{ at bat})] \cdots [1 - P(\text{hit on } 4^{th} \text{ at bat})]$

$= \quad 1 - [1 - .303]^4 = .764.$

Compare the details of this solution with the general technique for finding $P(A$ or B or $C \ldots)$ when the events are not pairwise mutually exclusive but are independent.

(ii) For the solution to part (ii) we assume independence and apply the multiplication rule:

$$P(\text{at least one hit in each of 44 consecutive games})$$

$= \quad P(\text{at least one hit in game 1 and}$
$\qquad \text{at least one hit in game 2 and} \ldots)$

$= \quad P(\text{at least one hit in game 1})$
$\qquad \cdot P(\text{at least one hit in game 2}) \cdots$

$= \quad (.764)(.764) \cdots = (.764)^{44} = .000007179.$

Note: The probability that he would get at least one hit in a consecutive string of 44 games at some time during the season is larger than the answer in (ii) but much more difficult to determine.

Example 2.14. We return to another variation of the birthday problem. Three sisters who live near Provo, Utah all gave birth on March 11, 1998. This is obviously a rare event. How rare? This raises the question, "What is the probability that three sisters will give birth on the same day?"

Before we even attempt to answer this question, there are two issues to consider. The first issue deals with the interpretation of the question. What does the question mean? Are there several different interpretations? The second issue deals with the assumptions built into the model we use to answer the question.

We focus on three of the possible interpretations of the question.

1. If each of three sisters will give birth in a given year, what is the probability that all three will give birth on March 11?

2. If each of three sisters will give birth in a given year, what is the probability that all three will give birth on the same day?

3. What is the probability that somewhere in the United States, there are three sisters who will give birth on the same day sometime in a given year?

The model we will use to answer these questions rests on two assumptions. One assumption is that all 365 days in a non-leap year are equally likely to be a birthday. The other assumption is that the day that one sister gives birth is independent of the days the other sisters give birth. These are basically the same assumptions that we used in Examples 2.5 and 2.6, but the assumptions are more questionable here because we are dealing with sisters rather than randomly selected women.

Let's look at the solution to each of the three possible interpretations:

1. P(all three sisters will give birth on March 11)

$$= \quad P(\text{1st sister gives birth on March 11, and 2nd sister gives birth on March 11, and 3rd sister gives birth on March 11})$$

$$= \quad \left(\frac{1}{365}\right)^3 = \frac{1}{48,627,125}.$$

2. P(all three sisters will give birth on the same day)

$$= \quad P(\text{1st sister gives birth on any day, and 2nd sister gives birth on same day as 1st sister, and 3rd sister gives birth on same day as 1st sister})$$

$$= \quad \left(\frac{365}{365}\right)\left(\frac{1}{365}\right)^2 = \left(\frac{1}{365}\right)^2 = \frac{1}{133,225}.$$

3. Suppose there are a total of n groups of three sisters in the United States who will give birth in a given year.

P(one group of three sisters will give birth on the same day)

$$= \quad \left(\frac{365}{365}\right)\left(\frac{1}{365}\right)^2 = \left(\frac{1}{365}\right)^2.$$

P(one group of three sisters will not give birth on the same day)

$$= \quad 1 - \left(\frac{1}{365}\right)^2.$$

P(none of the n groups of three sisters will give birth on the same day)

$$= \quad \left[1 - \left(\frac{1}{365}\right)^2\right]^n.$$

P(at least one group of three sisters will give birth on the same day)

$$= \quad 1 - \left[1 - \left(\frac{1}{365}\right)^2\right]^n.$$

The value of n can be estimated from data analytic procedures applied to health statistics. These procedures are outside the scope of this text. For additional information see reference [64]. Using these procedures, it can be shown that

$$P(\text{at least one group of three sisters will give birth on the same day}) > \frac{1}{407}.$$

2.5   Problems

1. Use the multiplication rule to find the probability of getting three different faces when three balanced dice are rolled. Compare your answer with that of Problem 8 in Chapter 1.

2. Answer the question in Example 1.5 in Chapter 1 using the multiplication rule.

3. Refer back to Example 1.11. The professor teaching this course is hoping that he won't have to give away any prizes. Suppose there are 65 students in class the day of the lottery. What is the probability that he will not have to give away any prizes.

4. Three balls are tossed into five boxes at random (each ball is equally likely to fall into each box).

   (i) Find P(three balls fall into different boxes).

   (ii) Find P(some box contains more than one ball).

   (iii) Rephrase Problem 1 in Chapter 2 into a problem which involves tossing balls into boxes.

5. A family has three children. Assume boys and girls are equally likely. Find the probability that the youngest child is a boy given that at least two of the children are boys.

6. Two William and Mary students arrive late for the math final exam and give the professor the phony excuse that their Mercedes Benz had a flat tire. "Well," the professor said, "each one of you write down on a piece of paper which of the four tires was flat." What is the probability that both students pick the same tire? Use the multiplication rule.

7. In a restaurant where a William and Mary student works, there are three levels—the downstairs cellar which seats 100 people; the main floor which seats 100 people; and the upstairs which seats 50 people. A customer being seated in the downstairs cellar complains that this is the fourth time she has been at this restaurant and each time she has been seated in the cellar. If seatings are random, what is the probability of this event?

8. Seven students take a probability test in a row of desks, with an aisle at each end of the row. They leave when they finish the test

in a random order. The instructor notices that all the students are able to go to one of the two aisles without passing by another student still taking the test. He wonders what the probability is of this happening. What is it?

9. If there are 253 people in a room find the probability that at least one person has a birthday on October 23. Discuss your answer in terms of the statement in the last paragraph before Example 2.6.

10. A student gives the following incorrect answer to Problem 9 in Chapter 2: $\left(\frac{1}{365}\right)^{253}$. State a question related to that in Problem 9 in Chapter 2 for which this is the correct answer.

11. If there are 5 people in a room find the probability that at least two people have the same birthmonth. (Assume all 12 birthmonths are equally likely.)

12. A student attempts to solve Example 2.6 by reasoning as follows:

P(at least one audience member has a birthday on October 23)

$=$ P(1st member has birthday on October 23

or 2nd member has birthday on October 23 or ...)

$=$ P(1st member has birthday on October 23)

$+ P$(2nd member has birthday on October 23) $+ \ldots$

$= \dfrac{1}{365} + \dfrac{1}{365} + \dfrac{1}{365} + \cdots + \dfrac{1}{365} = \dfrac{120}{365}.$

Is this solution correct? Why?

13. Describe how to use the random number target to carry out a simulation to estimate the probability in Problem 11 of Chapter 2. Do not actually carry out the simulation.

    Hint: Since there are 12 months and the target only has 6 numbers, you need to take a pair of numbers from the target for each simulation.

14. Assume all seven days of the week are equally likely to be a birthday. Suppose there are four people in a room. Find the probability that:

    (i) No two people in the room were born on the same day of the week;

(ii) Two or more people in the room were born on the same day of the week;

(iii) At least one person was born on a weekday (Monday through Friday).

15. A fair die is rolled six times. If the number of the roll and the number on the die agree, we say that a match has occurred. For example, if the first roll produces a one, then a match has occurred. If the second roll produces a two, then a match has occurred, etc. What is the probability that at least one match occurs?

16. Five balanced dice are rolled. Find the following probabilities:

(i) All 5 numbers are different.

(ii) At least two dice show the same number.

(iii) At least one die shows a 6.

17. A box contains four chips numbered 1,2,3,4. Two chips are drawn, at random, without replacement from the box. Let A = "sum of the numbers drawn is even" and B = "one of the numbers drawn is 4."

(i) List the elements in S.

(ii) List the elements in A.

(iii) List the elements in B.

(iv) Are A, B mutually exclusive events?

(v) Does $P(A \text{ or } B) = P(A) + P(B)$ here? Verify your answer numerically. Does your answer make sense in terms of your answer to part (iv)?

(vi) List the element(s) in "A and B."

(vii) Does $P(A \text{ and } B) = P(A)P(B)$? Are A, B independent events?

18. Nine children are seated at random in three rows of desks. Each row has three desks. Let A = "Al sits in the front row." Let B = "Al sits at one of the four corner desks."

(i) Are A and B independent events? Justify.

(ii) Are A and B mutually exclusive events? Justify.

19. In Problem 17, Chapter 2, find each of the following probabilities by reducing the sample space (without finding the event "A and B"):

   (i) $P(A|B)$;

   (ii) $P(A|\text{not } B)$;

   (iii) Are your answers to (i) and (ii) the same? Why?

20. Circle the correct letter and explain your answer. In an ancient matching game, two players show their right hands to each other simultaneously, with one, two, or three fingers extended. Suppose each player is equally likely to show one, two, or three fingers.

   Let A = "total number of fingers showing is odd," B = "both players show the same number of fingers," C = "total number of fingers showing is 4."

   (i) A, B are mutually exclusive and B, C are independent.

   (ii) A, B are independent and B, C are mutually exclusive.

   (iii) A, B are independent and B, C are independent.

   (iv) A, B are mutually exclusive and B, C are not independent.

   (v) A, B are not independent and B, C are not independent.

21. Select a card from an ordinary deck of 52 cards. Let A = "ace" and B = "spade." Are A and B independent if:

   (i) All cards are equally likely;

   (ii) The ace of spades is marked and the dealer makes it turn up twice as often as the other cards.

22. On May 17, 1998, David Wells of the New York Yankees pitched a perfect game (no runs, no hits, no errors). The only other time in history that a Yankee pitcher tossed a perfect game was in the 1956 World Series pitched by Don Larsen. An interesting coincidence is that Wells and Larsen had both attended the same high school. Give an approximate probability for this occurrence. To simplify computations, assume that all high schools have the same number of students. If you are having difficulty solving the problem in general, try solving the problem if there were four high schools in the U.S. and each high school had a total of 100 students over the years. Then, solve the original question. Explain the differences and the similarities between this problem and Problem 6 in Chapter 2.

23. Suppose we have a baby monkey whose written language consists only of the three characters A, B, space. In addition, suppose

    $P(A) = 3/6;$ $P(B) = 2/6;$ $P(\text{space}) = 1/6;$

    $P(A|\text{space}) = 1/3;$ $P(B|\text{space}) = 2/3;$ $P(B|A) = 2/3;$

    $P(\text{space}|A) = 1/3;$ $P(A|B) = 5/6;$ $P(\text{space}|B) = 1/6.$

    (i) Use the random number target to generate four characters from a first-level baby monkey.

    (ii) Use the random number target to generate four characters from a second-level baby monkey. Note that AA, BB cannot occur.

24. Competitive swimmers generally prefer the two outer lanes of the pool. These lanes have less waves, a clearer view of the timer, and are influenced the most by encouraging cheers from teammates. If a competitive swimmer has four separate races in a six-lane pool, what is the probability that she gets an outer lane at least one time? Assume swimmers are assigned lanes at random and every lane has exactly one swimmer.

25. According to *The Insiders' Guide to North Carolina's Outer Banks*, hang gliding is a fairly safe sport though less safe than bowling. The probability of injury on any flight is .003. If over every spring break for the next four years I make fifty flights, what is the probability that I will suffer at least one injury every year?

26. As a followup to Example 2.13, show that Mickey Mantle who had a lifetime batting average of .298 would be approximately two-thirds as likely to get at least one hit in each of 44 consecutive games as Pete Rose.

27. Refer to Example 2.13. State the questions answered by the following probabilities:

    (i) $1 - \left[(1 - .303)^4\right]^{44}$

    (ii) $1 - \left[1 - (.303)^4\right]^{44}$

28. In a weekly football selection contest, each student has to pick the winner of 12 football games. Suppose that student knowledge is such that each student has probability .51 of getting each game correct. If all 5000 Butler University students independently pick 12 games on a fall weekend, what is the probability that at least one student gets all 12 correct?

29. On March 24, 1998 in Jonesboro, Arkansas, students and teachers evacuated Westside Middle School for a fire alarm. The alarm had been sounded by two armed boys who were waiting outside the school. The two armed boys fired into the crowd of students and teachers killing four female students and one female teacher. After the tragedy there was some speculation about why all those killed were females. On ABC's *Prime Time Live*, a parent stated that the victims were apparently "selected because of their sex or who they were. It was not a random shooting where you just shoot out there." Find the probability that all five victims were female if males and females were equally likely in the crowd.

30. A total of 1000 marriages took place in Pittsburgh last year. Find the probability that for at least one of these couples, both partners celebrated their birthday on the same day of the year.

31. The winning three-digit number in a state lottery is determined each day by a random selection of numbers between 000 and 999 inclusive. One day the winning three-digit number was the same as the winning three-digit number on the previous day. The newspapers made a big issue about it. So the state had the selection procedure checked, but it turned out to be in proper working order. The lottery had been running for four years. What is the probability that during those four years there was at least one time when one day's winning number was the same as the previous day's winning number?

32. Automatic garage door openers have a transmitter in the car and a receiver in the garage. There are settings on both the transmitter and the receiver. For the opener to operate, the car transmitter and garage receiver must have an identical setting. There are 256 matching settings that an owner can select. In a neighborhood with 11 residents, suppose each resident chooses a random setting for his automatic garage door opener. Find the probability that at least one resident will open a neighbor's garage door as well as his own.

33. Two years after the Massachusetts Numbers Game began operating on April 10, 1976, a newspaper article, reporting on it, commented that as was to be expected, none of the randomly selected four digit numbers had been repeated. Assume the game is played every day of the year. Find an expression for the probability of the stated event. Is this probability less than .001?

34. On the *Late Show with David Letterman*, a regular feature is the appearance of a woman named Deborah Lynn, who describes herself as "intuitive." On each appearance, Deborah visits the Hello Deli near the theatre where the show is filmed and attempts to intuit which one of the Deli's 14 sandwiches are placed in front of her while blindfolded. As of September 13, 2003, Deborah had been wrong on 12 straight attempts. If all 14 sandwiches are equally likely, find the probability of Deborah being wrong on 12 straight attempts if she is guessing.

35. A *palindrome* is a sequence of characters that reads the same forward or backward. For example, the words *madam* and *rotator* are palindromes. The number 7448447 is a 7-digit palindrome. What is the probability that a randomly selected 7-digit number is a palindrome?

36. On a campus of 1000 students, a student tells a rumor to a second student, who in turn repeats it to a third student, etc. At each step, the recipient of the rumor is chosen at random from the 999 students available. Find the probability that the rumor will be told 999 times without returning to the originator. If you are having difficulty with this question, try a smaller version of the problem first. For example, replace 1000 by 4 and 999 by 3.

37. The Cathedral of Learning at the University of Pittsburgh is the tallest educational building in the western world. There are 41 floors above the ground level. There are several elevators on the ground level where students enter the building. Elevator #1 stops at even-numbered floors only. Suppose eight students enter elevator #1 on the ground level and exit the elevator independently and equally likely at each of the 20 stops. Find the probability that elevator #1 stops at floor number 16.

38. York County, Virginia, mails out the real estate tax bill and car tax bill to residents at the same time. Each bill is sent out with an identically addressed return envelope. So each resident has a choice of sending back each payment in one envelope (two envelopes in total) or both payments in the same envelope (one envelope in total).

    Suppose $p = P$(an envelope gets lost in the mail).

In terms of p:

(i) Find P(both checks arrive at the county office) if the resident uses 1 envelope.

(ii) Find P(both checks arrive at the county office) if the resident uses 2 envelopes.

How many envelopes should the resident use if:

(iii) She wants to maximize the probability that both checks arrive at the county office.

(iv) She wants to minimize the probability that both checks will be lost.

How many envelopes would you use? Ignore postage costs in your answer.

39. In Reference [5], the author states, "Fifty years ago, a family with an average number of children had a 50-50 chance of experiencing the death of a child." The author does not refer to a specific population. To try to gauge whether or not the statement sounds reasonable, we can by trial and error find numbers of children in a family and child death rates which produce this 50-50 chance.

For example:

(i) Suppose the child death rate is .21. Show that if there are 3 children in a family, then there is approximately a 50-50 chance that the family will experience the death of a child.

(ii) Suppose the child death rate is .13. Show that if there are 5 children in a family, then there is approximately a 50-50 chance that the family will experience the death of a child.

(iii) Suppose the child death rate is .10. Show that if there are 7 children in a family, then there is approximately a 50-50 chance that the family will experience the death of a child.

(iv) Check any source available to get a ballpark estimate of the child death rate and/or the number of children per family in 1950 and then decide if you think that the author's statement sounds reasonable.

40. The NCAA men's basketball tournament consists of 64 of the top college basketball teams competing to decide which one is the best

team in the country. The 64 teams are grouped into four divisions of 16 teams each. After the first round of the tournament in which eight of the teams are paired against the other eight teams in each division, the losing teams are eliminated from the tournament. After the second round of the tournament in which four of the remaining teams are paired against the other four remaining teams in each division, the losing teams are again eliminated from the tournament. In fact, the losing teams are eliminated after each round and therefore after four rounds only four teams are left in the tournament. There is one team left from each division. These remaining four teams are known as the "final four." Find the probability that a person will correctly predict all four of the original 64 teams which end up as the final four. Assume all teams are evenly matched.

41. The following is an excerpt from "Covering the Cops," by Calvin Trillin, which is a profile of Miami crime reporter, Edna Buchanan (see reference [51] in the Bibliography):

> In the newsroom of the Miami *Herald* there is some disagreement about which of Edna Buchanan's first paragraphs stands as the classic Edna lead. I line up with the fried-chicken faction. The fried-chicken story was about a rowdy ex-con named Gary Robinson, who late one Sunday night lurched drunkenly into a Church's outlet, shoved his way to the front of the line and ordered a three-piece box of fried chicken. Persuaded to wait his turn, he reached the counter again five or ten minutes later, only to be told that Church's had run out of fried chicken. The young woman at the counter suggested that he might like chicken nuggets instead. Robinson responded to the suggestion by slugging her in the head. That set off a chain of events that ended with Robinson's being shot dead by a security guard. Edna Buchanan covered the murder for the *Herald*—there are policemen in Miami who say it wouldn't be a murder without her—and her story began with what the fried-chicken faction still regards as the classic Edna lead: "Gary Robinson died hungry."

Later in the profile, we learn that there were 621 murders in Dade County, Florida, in 1981. Calvin Trillin continues by stating that a homicide detective told him that meant that a police reporter could drive to work in the morning knowing that there would almost certainly be at least one murder to write about. Verify whether or not this statement is true by answering the following question: If there are 621 murders in a year, what is the probability that there will be at least one murder on a particular day?

3

Combining the Addition and Multiplication Rules

3.1 Combining the Addition and Multiplication Rules

There are many cases where an event can be decomposed into its' components and the probability of this event can then be determined by applying the addition rule followed by the multiplication rule.

Example 3.1. Suppose the Mets and Cubs play a baseball doubleheader, and the Cubs are twice as likely to win any game as the Mets. Is a sweep more likely than a split?

Solution: Let's construct an urn model for this situation. Urn I will represent the first game and will contain 2 C's and 1 M since the Cubs are twice as likely to win game 1 as the Mets. Urn II will represent the second game and have the same composition as Urn I. We now select one letter at random from each urn to correspond to an outcome for the doubleheader. A sweep would correspond to getting the same letter twice.

Let A denote the event that the Cubs win game I and B denote the event that the Cubs win game II. How would we write the event "sweep" in terms of A, B?

First, observe that "not A and not B" = "Mets win doubleheader." Thus, we could write out the decomposition of the event "sweep" as follows:

$$\text{Sweep} = \text{(Cubs win doubleheader) or (Mets win doubleheader)}$$

$$= \text{(A and B) or (not A and not B).}$$

The two events connected by "or" are mutually exclusive so the addition rule applies giving:

$$P(\text{sweep}) = P(A \text{ and } B) + P(\text{not } A \text{ and not } B).$$

Now we have two events connected by the word "and" in each term above.

Here, we can assume that the events are independent. That is, whether or not a particular team wins the first game should not affect the probability of that team winning the second game (Using different urns implies A, B are independent since the letter selected from I would not affect the probability of getting a certain letter from II). Applying the multiplication rule for independent events, we get

$$P(\text{sweep}) = P(A)P(B) + P(\text{not } A)P(\text{not } B).$$

Substitution gives:

$$P(\text{sweep}) = \left(\frac{2}{3}\right)\left(\frac{2}{3}\right) + \left(\frac{1}{3}\right)\left(\frac{1}{3}\right) = \frac{5}{9}.$$

Since sweeps and splits are complementary events,

$$P(\text{split}) = 1 - P(\text{sweep}) = \frac{4}{9}.$$

Thus sweeps are more likely than splits for this example.

We could have solved this problem by enumeration using the methods of Chapter 1. In order to do so, we need to construct an equally probable sample space.

Consider the sample space $S = \{(C, C), (C, M), (M, C), (M, M)\}$ where the first letter in the ordered pair is the winner of game 1 and the second letter is the winner of game 2. This is not an equally probable sample space since (C, C) is more likely than (M, M). That is, the Cubs are more likely to win a doubleheader than the Mets since they are more likely to win each game.

We want the Cubs to be twice as likely to win any game as the Mets, so we could introduce $C1, C2, M$ as the three equally likely outcomes for each game. Then an equally probable sample space is

$$S = \{(C1, C1), (C1, C2), (C1, M), (C2, C1), (C2, C2),$$
$$(C2, M), (M, C1), (M, C2), (M, M)\},$$

where the letter in the first entry in the ordered pair is the winner of game 1, and the second entry letter is the winner of game 2.

In the equally probable sample space above, five of the outcomes in S (underlined) correspond to a sweep and so $P(\text{sweep}) = \frac{5}{9}$ which agrees with the answer we got using the addition and multiplication rules. In general, solution by enumeration is more tedious and sometimes not feasible since the sample space may be very large. By combining the addition and multiplication rules we have a more powerful technique.

An interesting observation about Example 3.1 is that it is possible to show that regardless of the probability of the Cubs winning any particular game, splits are never more likely than sweeps. The only time that sweeps and splits are equally likely is when the teams are evenly matched for each game. A check of the *Baseball Digest* for most years will reveal that a majority of major league doubleheaders ended in sweeps.

The remainder of this chapter deals with applications combining the addition and multiplication rules. A graphical technique for handling these applications is covered in Section 3.5 which is optional.

⋆ ## 3.2   Simpson's Paradox

This section deals with combining tables which can lead to counterintuitive results. The paradox was first discussed in a paper by E. H. Simpson which was published in 1951.

Example 3.2. Last semester at Albright College in a certain sophomore level math course the following data were observed:

	Total #	#A's	Chance of an A
Female Students	200	62	.31
Male Students	120	54	.45

Assuming no difference in abilities it appears there is a gender bias in grading. But the affirmative action official at the College observes the following breakdown by professor:

Professor I:

	Total #	#A's	Chance of an A
Female Students	40	22	.55
Male Students	100	50	.50

Professor II:

	Total #	#A's	Chance of an A
Female Students	160	40	.25
Male Students	20	4	.20

(Check the sums in the tables above.)

There is a greater overall chance of a male student making an A, but with each professor, females had a greater chance of getting an A. On an intuitive basis we see that a much higher percentage of female students enrolled in Professor II's class than in Professor I's class and Professor II is clearly a harder grader.

Let's look at the situation on a probabilistic basis. If

$$A = \text{``grade of A,''} \qquad F = \text{``female student,''} \qquad M = \text{``male student,''}$$
$$I = \text{``Professor I,''} \qquad II = \text{``Professor II,''}$$

then rewriting the above in terms of conditional probabilities

$$P(A|M) = .45 > P(A|F) = .31$$

even though the direction of the inequality is reversed for both males and females as seen below:

$$P(A|I \text{ and } M) \;=\; .50 < P(A|I \text{ and } F) = .55,$$
$$P(A|II \text{ and } M) \;=\; .20 < P(A|II \text{ and } F) = .25.$$

These three inequalities constitute Simpson's Paradox.

An important relationship is that the probabilities on each side of the three inequalities are related. For example $P(A|M)$ is a weighted average of $P(A|I \text{ and } M)$ and $P(A|II \text{ and } M)$.

To see that this is the case, observe that an A can be received from Professor I or from Professor II. Thus,

$$P(A) = P(I \text{ and } A \text{ or } II \text{ and } A).$$

The two events connected by the word "or" are mutually exclusive so the addition rule applies giving:

$$P(A) = P(I \text{ and } A) + P(II \text{ and } A).$$

Applying the multiplication rule we get

$$P(A) = P(I)P(A|I) + P(II)P(A|II).$$

Given a male student, this equation becomes:

$$P(A|M) = P(I|M)P(A|I \text{ and } M) + P(II|M)P(A|II \text{ and } M).$$

This equation shows that $P(A|M)$ is a weighted average of $P(A|I \text{ and } M)$ and $P(A|II \text{ and } M)$. The weights are $P(I|M)$ and $P(II|M)$. Substituting the appropriate values into the last equation to check numerically gives

$$.45 = \left(\frac{100}{120}\right)(.50) + \left(\frac{20}{120}\right)(.20).$$

Explain why $P(I|M) = \frac{100}{120}$. (Hint: How many male students are there in total? Of these, how many studied with Professor I?) The weights indicate that a high percentage of males enrolled in the easier class.

⋆ 3.3 Randomized Response Designs

Randomized response designs are surveys which use a probabilistic device (such as a random number target) to get information about some group of people regarding a sensitive question. Randomized response models are used in situations where the respondent might be uneasy about participating in the survey or might tend to give a false response because of the sensitive nature of the question.

Example 3.3. Suppose that we have a class of 100 college students and we want to estimate the proportion of students attending this college who have cheated on a high school exam. To accomplish this goal, the instructor asks each student in the class to select a number from the random number target in such a way that no one else can see his or her number. The instructor does this to protect confidentiality. The students are then told "Raise your hand if either you have cheated on a high school

exam or if you selected an odd number from the random number target."
A raised hand implies that either the student cheated on a high school
exam or got an odd number from the target or both (the two events
are not mutually exclusive). So a raised hand does not ensure that the
student has cheated. Thus the student is giving the instructor information
on a probabilistic basis yet confidentiality is maintained. Suppose that
77 of the 100 students raise their hands. We expect approximately 50
of these got an odd number from the target. The remaining 27 raised
hands suggest that 27 of the approximately 50 remaining students with
an even number had cheated. Thus we estimate that 27/50=54% of the
students cheated on a high school exam. So we have a method for making
an estimate about a population without knowing which students in the
sample actually cheated.

Several practical matters must be examined more closely before we
attempt to describe the solution to this example using rules of probability.
First, the sample of students from the college should be selected using a
random sample of all students from the college. One particular class might
be biased in some way. We presented Example 3.3 using one class solely
to make it easier to demonstrate how the model works. But in actually
carrying out such a study it is important to try to eliminate bias. As with
any survey we have sampling error since only a portion of the population
is being questioned. In our example, the population we wish to estimate
consists of all students attending the college. Second, there is error due to
randomization. Information is obtained on a probabilistic basis in order
to protect the confidentiality of the respondent. Because of these two
sources of error, randomized response surveys require a large sample size
to obtain a good estimate.

We now return to Example 3.3 to solve the problem using the rules of
probability.

Let p denote the true proportion of students at the college who have
cheated on a high school exam. Our objective is to estimate p based on the
information in the sample. What we know is that 77/100 or 77% of the
students in the class raised their hands. The technique for estimating p is
to set the given sample proportion equal to the corresponding probability.
Recall that a probability is a long-run proportion.

Thus, set

$$\text{Proportion of raised hands} = .77 = P(\text{raised hand}).$$

The next step is to find $P(\text{raised hand})$ in terms of p. This requires
a decomposition of the event "raised hand." A student might raise his

or her hand after selecting an even number or an odd number from the random number target. This gives us the decomposition for the event "raised hand."

P(raised hand) $= P$(odd number and raised hand or even number and raised hand).

The events connected by "or" are mutually exclusive so we use the addition rule to get:

P(raised hand) $= P$(odd number and raised hand)
$\qquad\qquad\qquad\quad + P$(even number and raised hand).

Applying the multiplication rule:

P(raised hand) $= P$(odd number)P(raised hand|odd number)
$\qquad\qquad\qquad\quad + P$(even number)$P$(raised hand|even number).

If a student obtains an odd number, he or she will always raise their hand. If a student obtains an even number, the student will raise his or her hand only if they have cheated on a high school exam. The probability that a randomly selected student had cheated is p. Substituting in the last equation gives:

$$P(\text{raised hand}) = \left(\frac{3}{6}\right)1 + \left(\frac{3}{6}\right)p = .77.$$

Solving the last equation for an estimate of p gives:

$$\left(\frac{3}{6}\right)p = .77 - \left(\frac{3}{6}\right)1 = .27.$$

So 54% is the estimate of p.

Once again, we have combined the addition and multiplication rules for our solution. This solution using the rules of probability is much preferred to the intuitive solution given earlier because it applies to a wider class of problems. One weakness of the technique used in Example 3.3 is that if the student did not raise his hand, the instructor knows for certain that the student did not cheat on a high school exam. In some sensitive questions, there might be a stigma attached to both a yes (raised hand) and no (hand not raised) response. One example would be to estimate the percentage of virgins at the college where there might be some stigma attached to either a yes or no answer for many respondents. The next example shows how we can use a variation of the technique used to solve Example 3.3 with another type of randomized response model so as to avoid this weakness.

Example 3.4. The honor council at Yale wants to estimate the proportion of students at Yale who have cheated on a college exam. (This proportion will be denoted by the letter p.) They select a sample of 500 students at random from the complete directory of students. Each student in the sample is handed a paper containing two statements:

> I have cheated on an examination at Yale (printed in red);
> I have never cheated on an examination at Yale (printed in black).

Each student then selects a card at random from a deck of cards which contains 26 red cards and 13 black cards. To preserve confidentiality, the interviewer does not see the card that is selected. The subject answers true (T) or false (F) to the statement corresponding to the color of the card that he has selected. That is, if a red card is selected, the subject will answer T if he or she has cheated and answer F otherwise. The interviewer records only T or F for each subject. When the subject gives a T response the interviewer does not know whether the response is to the first or second statement since the interviewer cannot see the card. Information is received but only on a probabilistic basis. If the deck had consisted of 26 red cards and 0 black cards this would correspond to the usual sampling situation where confidentiality is not protected. If the deck consisted of 26 red cards and 26 black cards no information is received by the interviewer. The composition of the deck used in this example of 26 red cards and 13 black cards gives information on a probabilistic basis. Exactly how the number 13 was chosen is beyond the scope of this book but it is related to the sample size needed to estimate p to within a certain degree of accuracy. From a practical point of view, if the deck were stacked too heavily in favor of either black or red cards, the respondent would lose faith in the procedure and be less inclined to answer honestly.

Suppose that of the 500 students responding, 209 give T answers. Recall from Example 3.3 that the procedure is to first set the sample proportion from the data equal to the corresponding probability since the sample is supposed to be representative of the population of students at Yale. Thus, set

$$P(T) = 209/500 = .418.$$

The next step is to find $P(T)$ in terms of p. This requires a decomposition of the event T. We could get a T response for either a red or a black card so the decomposition is:

$$T = \text{red card and } T \text{ or black card and } T;$$

$$P(T) = P(\text{red card and } T \text{ or black card and } T).$$

By the addition rule,

$$P(T) = P(\text{red card and } T) + P(\text{black card and } T).$$

By the multiplication rule,

$$P(T) = P(\text{red card})P(T|\text{red card}) + P(\text{black card})P(T|\text{black card}).$$

A student who answers T after selecting a red card has cheated and therefore $P(T|\text{red})=p$ since the proportion of cheaters is p. Now what is $P(T|\text{black})$? That is, of those who received black cards, what proportion answer T? The answer is the proportion who are not cheaters which is $(1-p)$.

Substituting in the equation we get from the multiplication rule gives

$$P(T) = \left(\frac{26}{39}\right)p + \left(\frac{13}{39}\right)(1-p).$$

Solving $P(T) = .418$ for the estimate of p gives

$$.418 = \left(\frac{26}{39}\right)p + \left(\frac{13}{39}\right)1 - \left(\frac{13}{39}\right)p$$

$$.418 = \left(\frac{13}{39}\right)p + \left(\frac{13}{39}\right).$$

The solution of this equation for p gives an estimate of .254. So we estimate that slightly over 25% of the students at Yale have cheated on an examination.

Surveys require a high response rate to produce unbiased results. Studies have shown that randomized response techniques give much higher response rates than anonymous questionnaires.

The following example uses the same reasoning involved in randomized response models but deals with a different application.

Example 3.5. (Probability in a Contested Election.) This example concerns the election for mayor in Flint, Michigan in 1975 between McCree(M) and Rutherford(R).

Recorded Vote

Precinct	M	R	Total
51	202	253	455
52	174	117	291
other	20099	20311	40410
Total	20475	20681	

The recorded vote shows R the winner by a margin of 206 votes. Precincts 51 and 52 had 5 and 4 voting booths respectively. The ballot assemblies were to be rotated between the two precincts, in the interest of fairness, so that the candidate whose name was first in Precinct 51 was second in Precinct 52. By mistake, the election officials placed one ballot assembly in Precinct 51 that should have been in Precinct 52 and one ballot assembly in Precinct 52 that should have been in Precinct 51. The result was that a voter using the booth with the wrong assembly in either precinct would have his vote recorded for the other candidate. The error in ballot assemblies was discovered after the polls had closed and machines disassembled making it impossible to distinguish which and how many votes were improperly recorded. We summarize the information in the following table:

	Vote Cast For	Vote Recorded For
Correct Assembly	M	M
Correct Assembly	R	R
Reversed Assembly	M	R
Reversed Assembly	R	M

Assume voters select booths independently and a booth with reversed ballot assembly is selected with probability equal to the reciprocal of the number of booths. So $P(\text{reversed assembly}) = \frac{1}{5}$ in Precinct 51 and $P(\text{reversed assembly}) = \frac{1}{4}$ in Precinct 52. We want to estimate the proportion of votes cast for M which we denote by the letter p. This will enable us to estimate the number of votes cast for M. First consider Precinct 51.

Set $P(\text{vote recorded for } M) = $ proportion of votes recorded for $M = \frac{202}{455}$.

Next find $P(\text{vote recorded for } M)$ in terms of p. To accomplish this, we decompose the event "vote recorded for M" by referring to the table above.

$$\begin{aligned}\text{Vote recorded for } M \ = \ & \text{correct assembly and vote cast for } M \\ & \text{or reversed assembly and vote cast for } R\end{aligned}$$

So,

> $P(\text{vote recorded for } M)$
>
> $\quad = \quad P(\text{correct assembly and vote cast for } M)$
>
> $\qquad + P(\text{reversed assembly and vote cast for } R)$

by the addition rule.

Applying the multiplication rule we get

> $P(\text{vote recorded for } M)$
>
> $\quad = \quad P(\text{correct assembly})P(\text{vote cast for } M|\text{correct assembly})$
>
> $\qquad + P(\text{reversed assembly})P(\text{vote cast for } R|\text{reversed assembly}).$

On substituting in the above equation, we get

$$P(\text{vote recorded for } M) = \left(\frac{4}{5}\right)p + \left(\frac{1}{5}\right)(1-p) = \frac{202}{455}.$$

Solving this equation gives
$$\left(\frac{3}{5}\right)p = \left(\frac{202}{455}\right) - \left(\frac{1}{5}\right) = \frac{111}{455}.$$

The estimate of p is
$$\left(\frac{111}{455}\right)\left(\frac{5}{3}\right) = \frac{37}{91}.$$

So the estimate of the number of votes cast for M in Precinct 51 is

$$\left(\frac{37}{91}\right)(455) = 185.$$

It follows that the estimate of the number of votes cast for R in Precinct 51 is
$$455 - 185 = 270.$$

Problems 17 and 18 at the end of this chapter ask that you estimate the number of votes cast for both candidates in Precinct 52 and then compare the estimated margin of victory with that which was recorded.

3.4   Bayes' Formula

Whenever we have a situation where we know $P(B|A)$ and we want to find the probability with the events reversed, $P(A|B)$, then Bayes' formula can be applied. The credit for the discovery of this formula is given to Thomas Bayes, a Presbyterian minister in England who was interested in mathematics. The derivation involves using the multiplication rule twice. The objective is to find $P(A|B)$. The first application of the multiplication rule gives

$$P(A|B) = \frac{P(A \text{ and } B)}{P(B)}.$$

The second application of the multiplication rule gives

$$P(A|B) = \frac{P(A)P(B|A)}{P(B)}.$$

The last equation is known as Bayes' formula. It gives $P(A|B)$ in terms of $P(B|A)$. To evaluate the denominator $P(B)$ in this equation, we will need to decompose B and then use both the addition and multiplication rules.

Keep in mind that there is a difference between $P(A|B)$ and $P(B|A)$. For example, if A denotes the set of college students and B denotes the set of single people, then $P(A|B)$ is the proportion of single people who are college students (less than .5) whereas $P(B|A)$ is the proportion of college students who are single (much larger than .5).

Consider another example of the difference between $P(A|B)$ and $P(B|A)$. Suppose that one in one million people have a certain DNA profile. This represents the probability of a DNA profile of suspect and crime sample match given that a suspect is innocent. But this is different from the probability with the events reversed. That is the probability that the suspect is innocent given a DNA match.

The next example illustrates how to apply Bayes' formula.

Example 3.6. Assume a screening test identifies 90% of the people who have the disease, the disease gives one false positive on the average out of 1000 patients, and the incidence rate for the disease is .0001. If the test indicates that the person has the disease, what is the probability that the person actually does have the disease?

Solution: Let

$$D = \text{"person has the disease" and } T = \text{"positive test."}$$

The first step is to translate the word problem into probability statements. Of the people who have the disease, the test is positive 90% of the time. Thus

$$P(T|D) = .90,$$

because we are reducing the sample space to those who have the disease. The incidence rate for the disease is .0001 so that $P(D) = .0001$. The disease gives one false positive on the average out of 1000 patients meaning this is how frequently there is a positive test when the person is free of the disease. Thus,

$$P(T|\text{not } D) = \frac{1}{1000} = .001.$$

The question asks, "If the test indicates that the person has the disease, what is the probability that the person actually does have the disease." So, find $P(D|T)$.

The objective is to find $P(D|T)$ when $P(T|D)$ is known. This requires the use of Bayes' formula.

$$P(D|T) = \frac{P(D)P(T|D)}{P(T)}.$$

Both quantities in the numerator are known so we focus on the denominator. The technique is to decompose the event in parentheses in the denominator.

$$T = (D \text{ and } T) \text{ or } (\text{not } D \text{ and } T).$$

That is, a positive test can occur when a person has the disease or when a person does not have the disease.

By the addition rule, $P(T) = P(D \text{ and } T) + P(\text{not } D \text{ and } T)$ since "D" and "not D" are mutually exclusive.

Using the multiplication rule we get

$$P(T) = P(D)P(T|D) + P(\text{not } D)P(T|\text{not } D).$$

Substituting the numerical values in the last equation gives

$$P(T) = (.0001)(.90) + (1 - .0001)(.001) = .00109.$$

This becomes the denominator in Bayes' formula giving

$$P(D|T) = \frac{P(D)P(T|D)}{P(T)} = \frac{(.0001)(.90)}{.00109} = .0826.$$

Only eight percent of those identified as having the disease actually do. The answer obviously depends heavily on the incidence rate. There are questions about the merits of large scale screening programs which are directed at diseases with low incidence rates especially when the diagnostic procedures involved may themselves be a health hazard. As this example indicates, a small percentage of those who test positive may actually have the disease in question.

Example 3.7. A cab was involved in a hit-and-run accident at night. Two cab companies, the Green and the Blue, operate in the city. Given the following information:

(a) 85% of the cabs in the city are Green and 15% are Blue.

(b) A witness identified the hit-and-run cab as Blue. The court tested the reliability of the witness under the same circumstances that existed on the night of the accident and concluded that the witness correctly identified each one of the two colors 80% of the time and incorrectly identified each one of the two colors 20% of the time.

What is the probability that the cab involved in the accident was Blue?

Solution: Let B denote the event that a cab is Blue, G denote the event that a cab is Green, and W denote the event that the witness identified the cab as Blue.

The event W is an observation just as the event T was an observation in Example 3.6. The events B, G are natural states which correspond to the events D, not D in Example 3.6.

From the information given in the problem, we see that $P(G) = .85$ and $P(B) = .15$. Since the witness correctly identified a Blue cab 80% of the time, $P(W|B) = .80$. The witness incorrectly identified a Green cab 20% of the time, so $P(W|G) = .20$. The objective is to find the probability that the cab involved in the accident was Blue, given that the witness identified the hit-and-run cab as Blue. That is, find $P(B|W)$.

Therefore $P(W|B)$ is known and the objective is to find $P(B|W)$. We need Bayes' formula,

$$P(B|W) = \frac{P(B)P(W|B)}{P(W)}.$$

Decomposing the event W in the denominator gives $W = (B$ and W or G and $W)$.

Therefore, $P(W) = P(B)P(W|B) + P(G)P(W|G)$ using the addition and multiplication rules.

Substituting numerical values into the last equation gives

$$P(W) = (.15)(.80) + (.85)(.20) = .29.$$

Substituting this value into Bayes' formula gives

$$P(B|W) = \frac{P(B)P(W|B)}{P(W)} = \frac{(.15)(.80)}{(.29)} = .414.$$

Two comments are in order.

1. The probability that the hit-and-run cab is Blue has increased from .15, in the absence of any evidence, to .414 with the report of the witness.

2. Note that $P(G|W) = 1 - P(B|W) = 1 - .414 = .586$. Therefore, $P(G|W) > P(B|W)$. This tells us that even with the report of the witness that the hit-and-run cab was Blue, the hit-and-run cab is more likely to be Green than Blue. This is because the credibility of the witness is not enough to overcome the large difference in relative frequencies between Blue and Green cabs.

Example 3.8. (Capture-Recapture Problem.) The Virginia Department of Fisheries wants to estimate the number of bass in a small pond. Suppose the pond is known to contain either 1, 2, or 3 bass with equal probabilities. We catch one bass at random, tag it, and release it. The next day we catch a bass at random from the pond and it is not tagged. Now, what probabilities should be assigned to the events that the pond contains 1,2,3 bass respectively?

Solution: Let B_1 denote the event that the pond contains one bass, B_2 denote the event that the pond contains two bass, and B_3 denote the event that the pond contains three bass. Let U denote the event that the bass we catch is untagged. We are given that

$$P(B_1) = P(B_2) = P(B_3) = \frac{1}{3}.$$

These are the prior probabilities that do not take into account that the event U has occurred. We want to find $P(B_1|U)$, $P(B_2|U)$, and $P(B_3|U)$. These are called the posterior probabilities as they take into account the additional information that the bass we catch is untagged. So, the idea is to adjust our prior probabilities to take into account the additional information that U has occurred.

If the bass we catch is untagged, it is not possible that there was only one bass in the pond since that bass would have been tagged. Thus, $P(B_1|U) = 0$.

To find $P(B_2|U)$, observe that $P(U|B_2)$ is known so Bayes' formula applies. Using Bayes' formula, $P(B_2|U) = P(B_2)P(U|B_2)/P(U)$.

To find the denominator we combine the addition and multiplication rules giving

$$P(U) = P(B_1 \text{ and } U) + P(B_2 \text{ and } U) + P(B_3 \text{ and } U);$$
$$P(U) = P(B_1)P(U|B_1) + P(B_2)P(U|B_2) + P(B_3)P(U|B_3).$$

If there are two bass in the pond, then we have a 50-50 chance of catching the untagged bass so $P(U|B_2) = \frac{1}{2}$. If there are three bass in the pond, then two of the three bass are untagged and $P(U|B_3) = \frac{2}{3}$. Substituting into $P(U)$:

$$P(U) = \frac{1}{3} \cdot 0 + \frac{1}{3} \cdot \frac{1}{2} + \frac{1}{3} \cdot \frac{2}{3} = \frac{7}{18}.$$

Thus,

$$P(B_2|U) = P(B_2)P(U|B_2)/P(U) = \frac{\left(\frac{1}{3}\right) \cdot \left(\frac{1}{2}\right)}{\left(\frac{7}{18}\right)} = \frac{3}{7}.$$

Similarly,

$$P(B_3|U) = P(B_3)P(U|B_3)/P(U) = \frac{\left(\frac{1}{3}\right) \cdot \left(\frac{2}{3}\right)}{\left(\frac{7}{18}\right)} = \frac{4}{7}.$$

So the additional information of catching an untagged bass makes it most likely that there are three bass in the lake, a little less likely that there are two bass in the lake, and impossible that there is only one bass in the lake.

A graphical technique for solving Bayes' formula problems is covered in Section 3.5 which is optional.

⋆ 3.5 Trees

A graphical method for solving problems in this chapter uses a horizontal tree.

Example 3.9. A probability class has two girls (G) and four boys (B). Two students are selected at random without replacement. Find $P(\text{2nd student selected is a girl}) = P(\text{2nd is } G)$.

Using the method of Section 3.1, we first decompose the event "2nd is G":

"2nd is G" = "1st is G and 2nd is G or 1st is B and 2nd is G."

So

$$
\begin{aligned}
P(\text{2nd is } G) &= P(\text{1st is } G \text{ and 2nd is } G \text{ or 1st is } B \text{ and 2nd is } G) \\
&= P(\text{1st is } G \text{ and 2nd is } G) + P(\text{1st is } B \text{ and 2nd is } G) \\
&= P(\text{1st is } G)P(\text{2nd is } G | \text{1st is } G) \\
&\quad + P(\text{1st is } B)P(\text{2nd is } G | \text{1st is } B) \\
&= \frac{2}{6} \cdot \frac{1}{5} + \frac{4}{6} \cdot \frac{2}{5} = \frac{2}{6}.
\end{aligned}
$$

An interesting thing to note here is that $P(\text{1st is } G) = P(\text{2nd is } G)$ even though the selections are made without replacement. At first sight this seems surprising. But $P(\text{2nd is } G) = \frac{2}{6}$ is an unconditional probability referring to the long-run proportion of times that the 2nd is a G whereas $P(\text{2nd is } G | \text{1st is } G) = \frac{1}{5}$ is a conditional probability referring to the long-run proportion of times the 2nd is a G but looking only at those cases where the 1st is a G. $P(\text{2nd is } G)$ considers the situation where the first student has been selected but no information regarding the sex of that student has been determined. So it is the same as if the first student had not been selected.

Next we find $P(\text{2nd is } G)$ using a (horizontal) tree:

1st selection   2nd selection

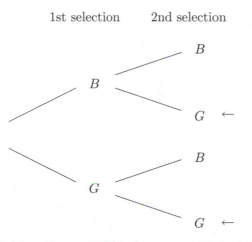

The horizontal tree gives a graphical representation of all possible outcome paths. In this problem there are four possible outcome paths where each path represents one of the mutually exclusive and exhaustive outcomes (B, B), (B, G), (G, B), and (G, G). The tree partitions the sample space for the experiment "select two students without replacement"

into all possible outcomes. Note that this is not an equally probable sample space as B's are more likely than G's. The two arrowed paths on the tree correspond to the event "2nd is G." To find the probability of this event using the tree, we begin by placing the appropriate probability notation on each branch. Each symbol represents the probability of going between outcomes on the tree.

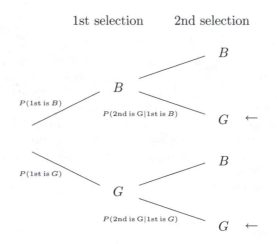

Multiplying horizontally on the two arrowed paths of interest use the multiplication rule to give the probabilities associated with these two paths.

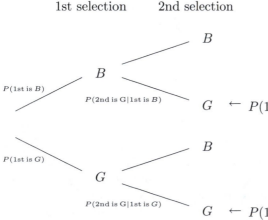

Using the addition rule, we add the two probabilities on the arrowed paths to get

$$P(\text{2nd is } G) = P(\text{1st is } B \text{ and 2nd is } G) + P(\text{1st is } G \text{ and 2nd is } G).$$

So the multiplication rule (for "and") is used when moving horizontally on the tree and the addition rule (for "or") is used when moving vertically.

Lastly, we insert the appropriate values in place of probability notation on the tree and carry out the computations.

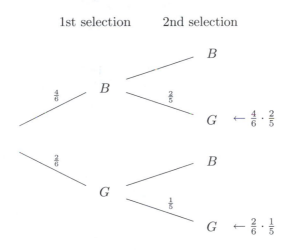

So, $P(\text{2nd is } G) = \frac{2}{6} \cdot \frac{1}{5} + \frac{4}{6} \cdot \frac{2}{5} = \frac{2}{6}$.

In solving a problem using the tree technique, it is only necessary to write down the last tree which contains the actual probability values. But one should keep in mind what those numbers represent and how the addition and multiplication rules are being represented graphically on a tree.

Example 3.10. Suppose the Mets and Cubs play a baseball doubleheader and the Cubs are twice as likely to win any game as the Mets. Find $P(\text{sweep})$. This is the solution to Example 3.1 using a tree.

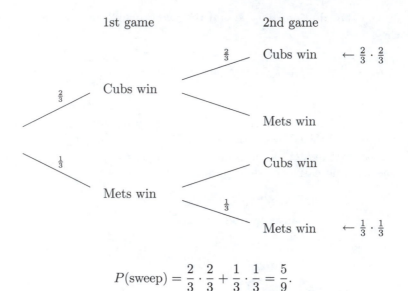

1st game   2nd game

$$P(\text{sweep}) = \frac{2}{3} \cdot \frac{2}{3} + \frac{1}{3} \cdot \frac{1}{3} = \frac{5}{9}.$$

Example 3.11. Refer back to Example 3.9. A probability class has two girls (G) and four boys (B). Two students are selected at random without replacement. Find $P(\text{1st is } G | \text{2nd is } G)$. Since $P(\text{2nd is } G | \text{1st is } G)$ is known, this is a Bayes' formula problem. Look at the tree.

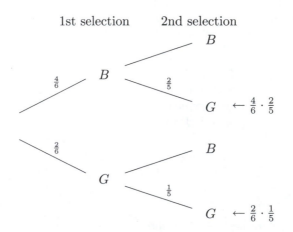

1st selection   2nd selection

$P(\text{1st is } G | \text{2nd is } G)$ is a conditional probability where the event which is given refers to the two arrowed paths on the tree. The question is: Given one of the two arrowed paths, what is the probability of the bottom arrowed path? This is the weight of the bottom arrowed path relative to the combined weight of the two arrowed paths. So the answer should be:

$$\frac{\frac{2}{6} \cdot \frac{1}{5}}{\frac{2}{6} \cdot \frac{1}{5} + \frac{4}{6} \cdot \frac{2}{5}} = \frac{1}{5}.$$

Let's verify that this is correct by applying Bayes' formula:

$$
\begin{aligned}
P(\text{1st is } G | \text{2nd is } G) &= P(\text{1st is } G \text{ and 2nd is } G)/P(\text{2nd is } G) \\[1em]
&= P(\text{1st is } G)P(\text{2nd is } G | \text{1st is } G)/P(\text{2nd is } G) \\[1em]
&= \frac{\frac{2}{6} \cdot \frac{1}{5}}{\frac{2}{6} \cdot \frac{1}{5} + \frac{4}{6} \cdot \frac{2}{5}} = \frac{1}{5}.
\end{aligned}
$$

We see that trees are a good visual tool for solving problems which combine the addition and multiplication rules.

⋆ 3.6   Applications in Cryptology

Cryptology is the science which deals with the coding and decoding of secret messages. Probability plays an important role in cryptology. In this section, we examine some fundamental concepts of cryptology.

In English-language text, the 26 letters in the alphabet appear with the following relative frequencies:

A	7.3%	N	7.8%
B	0.9%	O	7.4%
C	3.0%	P	2.7%
D	4.4%	Q	0.3%
E	13.0%	R	7.7%
F	2.8%	S	6.3%
G	1.6%	T	9.3%
H	3.5%	U	2.7%
I	7.4%	V	1.3%
J	0.2%	W	1.6%
K	0.3%	X	0.5%
L	3.5%	Y	1.9%
M	2.5%	Z	0.1%

We see that some letters occur much more frequently than others. The letter E occurs more frequently than the other letters. Suppose that two letters are selected from English-language text. From the relative frequency listing,

$$P(\text{both letters are } A\text{'s}) = (.073)^2,$$
$$P(\text{both letters are } B\text{'s}) = (.009)^2, \text{ etc.}$$

$P(\text{both letters are the same})$
$$= P(\text{both } A\text{'s or both } B\text{'s or } \ldots \text{ or both } Z\text{'s})$$
$$= P(\text{both } A\text{'s}) + P(\text{both } B\text{'s}) + \cdots + P(\text{both } Z\text{'s})$$
$$= (.073)^2 + (.009)^2 + \cdots + (.001)^2$$
$$= .066.$$

Now suppose we have a language with the same 26 letters but where all 26 letters are equally likely. Select two letters at random from that language.

$P(\text{both letters are the same})$
$$= P(\text{both } A\text{'s}) + P(\text{both } B\text{'s}) + \cdots + P(\text{both } Z\text{'s})$$
$$= \left(\frac{1}{26}\right)^2 + \left(\frac{1}{26}\right)^2 + \cdots + \left(\frac{1}{26}\right)^2$$
$$= 26\left(\frac{1}{26}\right)^2$$
$$= .038.$$

Summarizing,

$$P(2 \text{ letters are the same}) = \begin{cases} .066 & \text{for English language;} \\ .038 & \text{for 26 equally likely letters.} \end{cases}$$

We can estimate $P(2 \text{ letters are the same})$ for a given message by looking at the corresponding proportion. This is the same method we use to estimate the probability of a head in coin-tossing with the proportion of heads we get when we toss that coin. Here, the estimate of $P(2 \text{ letters}$ are the same) is equal to:

Proportion of pairs of letters in the message where the 2 letters are the same

$$= \frac{\text{number of pairs of letters in the message where 2 letters are the same}}{\text{number of pairs of letters in the message}}.$$

This estimate is called the "Index of Coincidence" or I.C. for the message.

Example 3.12. (Additive Shift.) Consider the following coded message which contains 11 letters (we use blocks of 5 letters to help disguise the message):

$$VLRAF \ OQVOX \ Q.$$

We begin our study of this message by calculating the Index of Coincidence (I.C.). To help find the correct number of pairs of letters, we use subscripts when the same letter is repeated.

$$V_1 LRAF \ O_1 Q_1 V_2 O_2 \ X \ Q_2.$$

Some ordered pairs of letters in the coded message are:

$$(V_1, L), (V_1, R), (V_1, A), (V_1, F), (V_1, O_1), (V_1, Q_1), (V_1, V_2), \ldots$$

There are $11 \times 10 = 110$ ordered pairs in total because there are eleven possibilities for the first letter in the pair and any of the ten remaining letters possible for the second letter in the ordered pair. Of these 110 ordered pairs, 6 ordered pairs have the same 2 letters. They are:

$$2 \times 1 \ = \ 2 \text{ pairs of } V\text{'s}: \ (V_1, V_2), \quad (V_2, V_1).$$

$$2 \times 1 \ = \ 2 \text{ pairs of } O\text{'s}: \ (O_1, O_2), \quad (O_2, O_1).$$

$$2 \times 1 \ = \ 2 \text{ pairs of } Q\text{'s}: \ (Q_1, Q_2), \quad (Q_2, Q_1).$$

Thus, I.C. $= 6/110 = .055$. It is difficult to draw conclusions based on such a short message. But since this estimate is closer to .066 than .038 it indicates that our coded message is slightly more consistent with an English-language message than a message from a language where the 26 letters are equally likely.

The coded message is given in the top line below. Each line below that corresponds to a shift of one letter. We continue to shift by one letter until we get a legible message. Here, by shifting 3 letters, we get the decoded message "You dirty rat."

V	L	R	A	F	O	Q	V	O	X	Q	← coded message
W	M	S	B	G	P	R	W	P	Y	R	
X	N	T	C	H	Q	S	X	Q	Z	S	
Y	O	U	D	I	R	T	Y	R	A	T	← decoded message

For the decoded message, I.C. is again .055. This quantity does not change in going from the coded to the decoded message—only the letters change. That is,

V in the coded message corresponds to Y in the decoded message.
O in the coded message corresponds to R in the decoded message.
Q in the coded message corresponds to T in the decoded message.

So in the decoded message, of the 110 ordered pairs of letters, the six ordered pairs that are the same are:

$$(Y_1, Y_2), (Y_2, Y_1), (R_1, R_2), (R_2, R_1), (T_1, T_2), (T_2, T_1).$$

For the additive shift, each letter in the alphabet of the coded message corresponds to a unique letter in the alphabet of the decoded message. In this example, the correspondence is as follows:

coded message alphabet   A B C D E F G H I J K L M

decoded message alphabet   D E F G H I J K L M N O P

coded message alphabet   N O P Q R S T U V W X Y Z

decoded message alphabet   Q R S T U V W X Y Z A B C

Because of the unique correspondence between the two alphabets, this coding system is not secure. That is, if a bad guy intercepts the coded message, it will be easy for him to decode. The next example gives us a method for making the coding system more secure. The sender transmits a secret key to the recipient.

Example 3.13. This time let's start out with the message we want to code:

$THEBU \ TLERD \ IDIT$ (The butler did it).

There are 14 letters in the message. Of these, there are 3 T's, 2 E's, 2 I's, and 2 D's. There are a total of $14 \times 13 = 182$ ordered pairs of letters in the message. Of these pairs,

$3 \times 2 \ = \ $ 6 pairs are (T, T), $2 \times 1 \ = \ $ 2 pairs are (E, E),

$2 \times 1 \ = \ $ 2 pairs are (D, D), $2 \times 1 \ = \ $ 2 pairs are (I, I).

So $6 + 2 + 2 + 2 = 12$ pairs of letters out of 182 have the same 2 letters. Here,

$$I.C. \ = \ 12/182 = .0659 \text{ which is near } .066$$

$$\text{indicating an English-language text.}$$

Now we make up a key and write that key repeatedly over the message. Let's decide on 1,5 as the key. Here the *length* of the key is 2 because the key has two numbers, but in general the length of the key could be any number greater than one.

1	5	1	5	1	5	1	5	1	5	1	5	1	5	
T	*H*	*E*	*B*	*U*	*T*	*L*	*E*	*R*	*D*	*I*	*D*	*I*	*T*	← message
S	*C*	*D*	*W*	*T*	*O*	*K*	*Z*	*Q*	*Y*	*H*	*Y*	*H*	*O*	← coded
														message

The coded message appears below the message. In Example 3.12, we found the message from the coded message. Here, we have to work backwards to find the coded message from the message. An additive shift of 1 letter is used to go from the coded message to the message for those letters headed 1 from the key. A shift of one letter changes S to T, D to E, etc. An additive shift of 5 letters is applied to letters in the coded message headed 5 from the key. A shift of 5 letters changes C to H, W to B, etc.

We see that there is no longer a unique correspondence between the coded and decoded message alphabets. The letter T can be coded into O or S depending on its position in the message. This has the effect of distributing the letters more equally in the coded message than in the message. So, in a long coded message using a longer key than used here, the letters will appear in nearly equal proportions.

In the coded message here, there are 2 Y's, 2 H's, and 2 O's. Of the total of 14 x 13 = 182 pairs of letters in the coded message, 2 pairs are (Y, Y), 2 pairs are (H, H), and 2 pairs are (O, O). Thus,

$$I.C. \quad = \quad (2 + 2 + 2)/182 = .033 \text{ which is more consistent}$$

with an equal distribution of letters.

This gives us some ideas about how to decode a message. We could calculate I.C. for the coded message. A result closer to .066 indicates a likely perfect correspondence between the coded and decoded message alphabets. A result closer to .038 indicates the message was coded using a key.

Exactly how to proceed if a key was used is beyond the scope of this book. But it makes sense to employ a trial and error technique with keys of different lengths until we find a key length where the I.C.'s corresponding to the letters for each number heading are near .066. That part of the message which is coded by the same number in the key will have the same letter frequency pattern as the English language.

Lastly, keep in mind that I.C. is a proportion based on a sample and it gives a good estimate of a probability only when the message is long.

3.7 Problems

1. Suppose the Mets and Cubs play a doubleheader. Find the probability of a Sweep:

   (i) if the Cubs are one and one-half times as likely to win any game as the Mets;

   (ii) if both teams are evenly matched.

2. Describe how to use the random number target to carry out a simulation to estimate P(Sweep) in Example 3.1. Do not actually carry out the simulation.

3. Two William and Mary students arrive late for the math final exam and give the professor the phony excuse that their Mercedes Benz had a flat tire. "Well," the professor said, "each one of you write down on a piece of paper which of the four tires was flat." What is the probability that both students pick the same tire? Use the addition and multiplication rules.

4. Draw two chips, at random and with replacement, from an urn containing chips numbered 1, 2, and 3. Let A = "sum of the digits on selected chips is an even number."

   We want to find P(A).

   (i) Solve the problem using the enumeration technique.

   (ii) Solve the problem using both the addition and multiplication rules.

   (Hint: Think about the decomposition of A. If the sum of two numbers is even, what do we know about whether the terms in the decomposition are odd or even?)

   (iii) Relate this problem to Example 3.1 by stating an event in this problem which corresponds to the following events in Example 3.1:

   (a) Sweep;

   (b) Mets win the doubleheader.

(iv) Would this problem change if we selected one chip at random from each of two urns each containing chips numbered 1, 2, 3?

5. If two evenly matched teams play a tripleheader (three games in one day), what is the probability of a sweep? (Major league baseball's last tripleheader was played between the Cincinnati Reds and the Pittsburgh Pirates on October 2, 1920.)

6. In the World Series, the first team to win four games wins the series. In the 1987 World Series, the St. Louis Cardinals won game 5 to give them a three-games-to-two edge over the Minnesota Twins. According to the Associated Press, "The numbers of history support the Cardinals and the momentum they carry. Whenever a World Series has been tied 2–2, the team that won game 5 was eventually the champion 71% of the time." If momentum is not a factor and each team has a 50% chance of winning each game, what is the probability that the Game 5 winner eventually wins the Series?

7. A drawer contains 3 blue socks and 8 white socks. Two of the socks are chosen at random with replacement.

   (i) What is the probability that both socks are the same color?

   (ii) Rephrase this problem in terms of an example involving baseball doubleheaders.

8. Amy and Barbara draw coins in turn, at random, from a bag containing two dimes and three nickels until the bag is empty. Amy draws first, then Barbara, then Amy, etc. It is known that Amy drew a dime before Barbara drew a dime. What is the probability that Amy drew a dime on the first draw? Do the problem two ways:

   (i) Listing all possible outcomes;

   (ii) Using a formula for conditional probability.

9. Jeannette gives the following solution to Problem 8 in Chapter 3. There are two possibilities for Amy drawing the first dime:

   (i) Amy draws a dime on the first draw;

   (ii) Amy draws a nickel, then Barbara draws a nickel, then Amy draws a dime.

Therefore the answer must be 1/2. Comment on Jeannette's solution.

*10. The following statistics for baseball players Dave Justice and Andy Van Slyke are for the 1989 and 1990 seasons. Does Simpson's Paradox apply?

	At bats	Hits	Average
1989			
Justice	51	12	.235
Van Slyke	476	113	.237
1990			
Justice	439	124	.282
Van Slyke	493	140	.284

*11. The question of whether or not the imposition of the death penalty is racially biased has been a recent topic in the news. In a paper from *American Sociological Review*, 46(1981), pp. 918–927 titled, "Racial characteristics and imposition of the death penalty," the following data appeared:

	White Defendant: Death Penalty		Black Defendant: Death Penalty	
	Yes	No	Yes	No
White Victim	19	132	11	52
Black Victim	0	9	6	97

Show that Simpson's Paradox holds: There is a greater chance of white defendants being sentenced to death overall, but for both black and white victims there is a greater chance that black defendants are sentenced to death.

*12. In Example 3.2, show that $P(A|F)$ is a weighted average of $P(A|I$ and $F)$ and $P(A|II$ and $F)$.

*13. The sensitive question is "Have you ever had an abortion?" The randomizing device is a box containing 10 red balls, 15 white balls, and 20 blue balls. The instructions are "Draw a ball at random from the box, making sure that no one else sees its color. If the ball is red, mark the answer yes. If the ball is white, mark the answer no. If the ball is blue, truthfully answer the stated question." In one sample of 125 women, there were 60 "yes" responses. Estimate the proportion of women in the population who have had an abortion.

*14. A researcher wishes to estimate the proportion p of university students who have ever cheated on an examination. She prepares a box containing 100 cards of which 20 contain question A and 80 contain question B.

> Question A: Is the last digit of your telephone number a 0, 1, or 2?

> Question B: Have you ever cheated on an examination?

Each student draws a card at random from the box and answers the question it contains. The student then returns the card to the box. Suppose that 18 of 100 students answer "yes." Give an estimate of p.

*15. Due to an incident in a classroom building at Bob Jones University, the administration has decided to conduct a survey to determine if students having sexual intercourse in academic buildings is a common occurrence. Three hundred students privately selected a ball from a box of balls numbered 1–20. The students are given a card with the following written on it:

1. I have had sex in an academic building.
2. I have never had sex in an academic building.

If the student picks a ball including the number 1 on it $(1, 10, 11 \ldots)$, he/she answers True or False to statement 1 on the card. If the student selects a ball with any of the other numbers on it, he/she answers True or False to statement 2.

If 136 of the students answer True, estimate the proportion of Bob Jones students who have had sex in an academic building.

*16. Estimate the proportion of people in the population who have committed adultery. Each subject in the sample draws a card at random from an ordinary deck of 52 cards and then responds with either a plus (+) or a minus (−) in accordance with the scheme below. The interviewer does **not** see the card selected.

During your Married Life, Have You Ever Had Sex with Someone other than Your Spouse?	Card Drawn	Response to Interviewer
Yes	Spade	+
Yes	Non-Spade	−
No	Spade	−
No	Non-Spade	+

A (+) response could mean that the person either did or did not have sex with someone other than his or her spouse. The same is true for the (−) response. Suppose that 81 out of 150 in the sample gave a (+) response.

*17. In Example 3.5, show that the estimate of the number of votes cast for M in Precinct 52 is 202.5 (remember that Precinct 52 had 4 booths rather than 5).

*18. Complete Example 3.5 by showing that the estimated margin of victory for R is 183 votes compared to the recorded margin of 206 votes.

19. Let A = "person is a U.S. senator" and B ="person is a U.S. male."

    Give ballpark estimates of the following probabilities based on your knowledge of the proportions rather than using Bayes' formula:

    (i) $P(A|B)$;

    (ii) $P(B|A)$.

    Hint: How many men are in the U.S. Senate?

20. In Example 3.6, are each of the following statements true or false?

    (i) $P(\text{not } T|D) = .10$.

    (ii) $P(D| \text{ not } T) = .9174$.

21. Suppose 20% of drivers on a stretch of Route 64 are speeding. The speed of each driver is estimated by radar at a speed trap. The radar catches 90% of those speeding, but it incorrectly indicates speeding for 15% of those driving within the speed limit. What are the chances that someone caught in the speedtrap is not guilty of speeding?

22. A student takes a multiple choice exam where each question has five possible answers. Some of the answers she knows, while others she gets right by making lucky guesses. Suppose that she knows the answer to 70% of the questions. What is the probability that she knew the answer to a question given that she got it right?

23. During a riot in Pittsburgh, 100 persons are arrested on suspicion of looting. Each is given a polygraph test. From past experience it is known that the polygraph is 90% reliable when administered to a guilty suspect and 98% reliable when given to someone who is innocent. Suppose that of the 100 persons taken into custody only 12 were actually involved in any wrongdoing. What is the probability that a suspect is innocent if the polygraph says she is guilty?

24. Recently a U.S. Senate Committee investigated the feasibility of setting up a national screening program to detect child abuse. A team of consultants estimated the following probabilities:

    (i) One child in 90 is abused;

    (ii) A physician can detect an abused child 90% of the time;

    (iii) A physician would incorrectly label 3% of all nonabused children as abused.

    What is the probability that a child is actually abused given that the physician diagnosed him as abused?

25. Two closely related species of mushrooms (I and II) are difficult to identify without the aid of a microscope. One method used in the field to separate the two species is to note the presence or absence of a ring on the stalk of the plant. Ninety percent of species I and 20% of species II have the ring. It is also known that in the particular area where the mushrooms are being studied, 70% of them are species I. Suppose the field worker finds a mushroom with a ring and decides it belongs to species I. What is the probability that he is correct?

26. In Example 3.8, suppose that the bass caught was tagged. Now, what probabilities would you assign to the events that the pond contains 1, 2, 3 bass respectively?

27. The Department of Wildlife needs to estimate the number of deer within a wildlife preserve. Suppose the preserve is known to contain

either 1, 2, 3, 4, or 5 deer with equal probabilities. We capture one deer at random, tag it, and release it. The next day we observe a deer from the preserve and it is not tagged. Now, what probability would you assign to the event that the preserve contains 3 deer?

28. In the 1982 contest for Governor of California, Bradley lost to Deukmejian by a margin of 10%. There were only two candidates. Suppose 40% of Bradley voters responded at the exit poll while 30% of Deukmejian voters responded. On the basis of exit polls, which candidate did the TV stations predict would win?

29. A writes to B and does not receive an answer. Assuming that one letter in ten is lost in the mail, find the probability that B received the letter. It is to be assumed that B would have answered the letter if he had received it.

30. If A = "defendant is innocent" and B = "DNA profile of defendant and crime sample match," fill in the blanks below with either $P(A|B)$ or $P(B|A)$:

   (i) Forensic evidence answers the probability _____ .

   (ii) Jury estimates the probability _____ .

*31. A researcher wishes to estimate the proportion p of university students who have ever cheated on an examination. He prepares a box containing 100 cards, 20 containing question A and 80 containing question B:

Question A: Were you born in July or August?
Question B: Have you ever cheated on an examination?

Each student draws a card at random from the box and answers the question it contains. Since only the student knows which question is being answered, confidentiality is assured, and the researcher expects all students to answer truthfully.

   (i) Suppose that 30 out of 100 students answer "yes." Give an estimate of p.

   (ii) Estimate the probability that a student who answers "yes" is replying to question B.

*32. The sensitive question is, "Have you ever had an abortion?" The randomizing device is a box containing 10 red balls and 15 white balls. The instructions are, "Draw a ball at random from the box, making sure that no one else sees its color. If the ball is red, mark the answer "yes." If the ball is white, truthfully answer the sensitive question. In one sample of 100 women there were 55 yes responses. Estimate the proportion of women in the population who have had an abortion.

*33. Instead of estimating the proportion of women who have had an abortion, we might want to estimate the proportion of women who have had 0, 1, 2, 3, etc., abortions. Suppose we have collected data in a study where we believe the maximum number of abortions is 5. Each female subject in our sample selects two numbers, in private, from the random number target. If the subject in our sample selects either 1 or 2 as her first number, she gives an answer to the interviewer which is determined by the second number she selects:

> If the second number selected is 1, 2, 3, 4, 5, she responds by giving that same number to the interviewer. If the second number selected is 6, she responds by giving the number 0 to the interviewer.

If the subject in our sample selects either 3, 4, 5, or 6 as her first number, she gives an answer to the interviewer which is determined by the actual number of abortions she has had:

> If she has had 0, 1, 2, 3, 4, or 5 abortions, she responds by giving the correct number of abortions to the interviewer (we assumed 5 was the maximum).

She just ignores the second number selected from the target. In this way, the interviewer cannot tell from a response of 3 whether the respondent has had 3 abortions or has selected the number 3 from the random number target after first selecting a 1 or 2.

The interviewer collected the following data:

Number of Subjects	Number Given to the Interviewer
900	0
600	1
400	2
200	3
150	4
150	5
2400	

Estimate the proportion of women in the population who have had 0, 1, 2, 3, 4, 5 abortions.

34. On the morning of October 2, 1982, the won-lost records of the three leading baseball teams in the western division of the National League were as follows:

Team	Won	Lost
Atlanta Braves	88	73
San Francisco Giants	87	74
Los Angeles Dodgers	87	74

Each team had one game remaining to be played. The San Francisco Giants had a game with the Los Angeles Dodgers and the Atlanta Braves game was with the San Diego Padres. Suppose that the outcomes of all remaining games are independent and each game is equally likely to be won by either team. What are the probabilities that each of the teams win the division? If two teams tie for first place, they have a playoff game which each team has an equal chance of winning. San Diego was not in contention to win the division.

35. A committee is made up of three members denoted by A, B, C. The committee makes a correct decision when the majority of the members on the committee make the correct decision.

   (i) The committee makes a correct decision when A makes a correct decision, B makes an incorrect decision, and C makes a correct decision. List the other three cases when the committee makes a correct decision.

   (ii) Suppose each member of the committee makes a decision independent of the other committee members. Suppose A makes a correct decision with probability .9, B makes a correct decision with probability .8, and C makes a correct decision with probability .7. Find the probability that the committee makes a correct decision.

   (iii) Suppose each member of the committee makes a decision independent of the other committee members. Suppose A makes a correct decision with probability .9, B makes a correct decision with probability .8, but C tosses a fair coin and makes a correct decision when the coin comes up heads. Find the probability that the committee makes a correct decision.

(iv) Suppose A and B make decisions independent of each other, but C always makes the same decision as B. Suppose A makes a correct decision with probability .9 and B makes a correct decision with probability .8. Find the probability that the committee makes a correct decision.

(v) Does the committee make a correct decision more often when C tosses a fair coin or when C makes the same decision as B?

36. Use the addition and multiplication rules to solve Problem 17 in Chapter 1.

37. The following article appeared in *USA TODAY* on June 3, 2002.

Greater Risk Earlier in Life

The American Cancer Society says protection against sun as a child is important because of the possible link between severe childhood sunburns and risk of melanoma later in life. Melanoma risk by age group:

Birth to age 39
  Male: 1 in 744
  Female   1 in 453

Ages 40 to 59
  Male: 1 in 190
  Female   1 in 249

Ages 60 to 79
  Male: 1 in 106
  Female   1 in 207

Source: "Cancer Facts & Figures, 2001," The American Cancer Society

(i) If $P(A|B) = 1/190$, describe the event A and the event B.

(ii) Write out an expression for P(male age 0 to 79 has melanoma).

(iii) What quantities must be known in order to evaluate the expression in (ii)?

*38. Solve Exercise 7(i) in Chapter 3 by using a tree.

*39. Solve Exercise 23 in Chapter 3 by using a tree.

*40. Solve Exercise 24 in Chapter 3 by using a tree.

*41. Suppose an alphabet has only 3 letters A, B, C. If two letters are selected, what is the probability that both letters are identical if:

   (i) The 3 letters are equally likely.

   (ii) A is twice as likely to occur as B which is twice as likely to occur as C.

*42. Suppose we have a message with 17 letters. Find P(at least one E appears) if we assume the letters occur independently and

   (i) All 26 letters are equally likely,

   (ii) We have English-language text.

You intercept a coded message with 17 letters which contains zero E's. Based on your answers to (i) and (ii) would you conclude the message was coded using a key?

*43. Given the coded message: EMGLB GYLQ.

   (i) Calculate the Index of Coincidence.

   (ii) Based on (i), do you think a key was used?

   (iii) Decode the message.

*44. Given the coded message: SEDPG XCLVH MLVP.

   (i) Calculate the Index of Coincidence.

   (ii) Do you think a key was used?

   (iii) If the key was 1,3 decode the message.

4

Random Variables, Distributions, and Expected Values

4.1 Random Variables, Distributions, and Expected Values

We are usually interested in a summary of outcomes rather than in the details of individual outcomes. If a baseball team plays 162 games in a season, we may prefer to know the total number of games they won rather than a complete listing of which games were won or lost. In this section we generalize and extend some previous results. We begin with two definitions.

Definition: A *random variable* associates a numerical value with each outcome of an experiment.

Definition: The values that a random variable assumes together with the corresponding probabilities is called the *distribution* of the random variable.

Example 4.1. The new commissioner of baseball wants to shorten the World Series so he decides that the first team to win two games will win the series. Suppose the series is between the Braves and the Red Sox, and the Braves are twice as likely to win each game as the Red Sox.

Let B = "Braves win a game" and R = "Red Sox win a game."

A sample space for this example would be

$$S = \{BB, RR, BRB, BRR, RBB, RBR\},$$

where the first letter refers to the first game, second letter to second game, etc.

Since the Braves are twice as likely to win a game as the Red Sox, $P(B) = \frac{2}{3}$, and $P(R) = \frac{1}{3}$. Note that S is not an equally probable sample space.

Let X = number of games in the series. Then X is an example of a random variable. For example, X associates the value 3 with the outcome RBB.

To find the distribution of X we need to find the probability that X takes on each possible value.

$$P(X = 2) = P(BB) + P(RR) = \frac{2}{3} \cdot \frac{2}{3} + \frac{1}{3} \cdot \frac{1}{3} = \frac{5}{9},$$

$$P(X = 3) = P(BRB) + P(BRR) + P(RBB) + P(RBR)$$

$$= \frac{2}{3} \cdot \frac{1}{3} \cdot \frac{2}{3} + \frac{2}{3} \cdot \frac{1}{3} \cdot \frac{1}{3} + \frac{1}{3} \cdot \frac{2}{3} \cdot \frac{2}{3} + \frac{1}{3} \cdot \frac{2}{3} \cdot \frac{1}{3} = \frac{4}{9}.$$

We write the distribution in table form as follows:

x	$P(X = x)$
2	$\frac{5}{9}$
3	$\frac{4}{9}$
	1

(We can provide a partial check on our work since the total of all the probabilities must equal 1.)

The lower case x represents a typical value which the random variable X assumes.

Definition: The *expected value* of a random variable X which is denoted by $E(X)$ is obtained by multiplying each value of the variable by its probability and then summing the products.

In Example 4.1,

$$E(X) = 2\left(\frac{5}{9}\right) + 3\left(\frac{4}{9}\right) = \frac{22}{9} = 2.444.$$

Note that $E(X)$ need not be an attainable value of the random variable X. In this example, 2.444 is not an attainable value of X. The definition tells us how to calculate an expected value but does not tell us what it means. The interpretation is that $E(X)$ is a long-run average in that if an experiment were repeated many times the average of the x values would approach $E(X)$. So $E(X)$ is a long-run average in the same way that probability is a long-run proportion. In Example 4.1, the average number of games in the series would approach 2.444 if the same teams played this series over and over again.

In Example 4.1, we were able to determine $E(X)$. Suppose we had not been able to do so. We could simulate the experiment a large number of times and then estimate $E(X)$. To demonstrate for Example 4.1, let the numbers 1, 2 from the random number target correspond to a Red Sox win and let 3, 4, 5, 6 correspond to a Braves win. By this assignment, the Braves will be twice as likely to win a game as the Red Sox. Now for each simulation continue to select numbers until the series is over. The following list of 20 simulations will enable us to estimate $E(X)$.

Simulation Number	Values from Random Number Target	Series Outcome	Number of Games
1	6, 1, 3	BRB	3
2	4, 5	BB	2
3	1, 6, 5	RBB	3
4	4, 2, 1	BRR	3
5	3, 3	BB	2
6	4, 4	BB	2
7	5, 4	BB	2
8	1, 1	RR	2
9	2, 2	RR	2
10	2, 3, 5	RBB	3
11	4, 1, 6	BRB	3
12	6, 5	BB	2
13	2, 2	RR	2
14	3, 5	BB	2
15	5, 6	BB	2
16	4, 1, 6	BRB	3
17	5, 6	BB	2
18	2, 4, 2	RBR	3
19	6, 2, 5	BRB	3
20	6, 3	BB	2

The estimate of $E(X)$ is the average value of the x's which is

$$(3 + 2 + 3 + 3 + 2 + \cdots + 3 + 3 + 2)/20 = 48/20 = 2.4.$$

As the number of simulations increases, the estimate approaches $E(X) = 2.444$.

Example 4.2. An article from the *New York Times* in May of 1990 entitled, "More in China Willingly Rear Just One Child," focused on an attempt by the Chinese Government to limit a family to one child. Although some in China had no problem with the policy because of practical economic reasons, others suggested revising the policy to limit families to one son. The "one son" policy states that as long as a woman gives birth to a girl she can have another child. But as soon as the woman gives birth to a boy, she can have no more children. This way, no family would have more than one boy, but plenty of families would have several girls. If the "one son" policy were adopted to reduce the growth of the Chinese population, several questions arise. Under the "one son" policy, what would be the average number of children in a family? Also, how would the "one son" policy affect the ratio of male to female births in China?

We focus on the solutions to this problem here and also later in Example 4.7. In modeling any real-life situation, some simplifying assumptions must be made. Here are three assumptions:

(i) Assume each birth is a single child, so there are no twins, triplets, etc., in the model.

(ii) Assume male and female births are equally likely.

(iii) Assume all couples are capable and desirous of having children.

Sometimes after a model is analyzed, the assumptions can be refined to see how adjusting some factors will affect the answers.

We add the additional condition here that there will be at most three children in the family. The restriction will be removed in Example 4.7.

Solution: Here, we introduce two random variables.

Let U = number of male(M) children in a family.
Let V = number of female(F) children in a family.

Assume $P(M) = P(F) = \frac{1}{2}$. Then $S = \{M, FM, FFM, FFF\}$, where the order of the letters represents the birth order. This sample space is not equally probable.

First we find the distribution of U and then $E(U)$:

$$
\begin{aligned}
P(U = 0) &= P(FFF) &= \tfrac{1}{2} \cdot \tfrac{1}{2} \cdot \tfrac{1}{2} &= .125. \\
P(U = 1) &= 1 - P(U = 0) &= 1 - .125 &= .875.
\end{aligned}
$$

The distribution of U is:

u	$P(U = u)$
0	.125
1	.875
	1

So $E(U) = 0(.125) + 1(.875) = .875$.

Next we find the distribution of V and then $E(V)$:

$$
\begin{aligned}
P(V = 0) &= P(M) &= \tfrac{1}{2} = .5. \\
P(V = 1) &= P(FM) &= \tfrac{1}{2} \cdot \tfrac{1}{2} = .25. \\
P(V = 2) &= P(FFM) &= \tfrac{1}{2} \cdot \tfrac{1}{2} \cdot \tfrac{1}{2} = .125. \\
P(V = 3) &= P(FFF) &= \tfrac{1}{2} \cdot \tfrac{1}{2} \cdot \tfrac{1}{2} = .125.
\end{aligned}
$$

The distribution of V is:

v	$P(V = v)$
0	.5
1	.25
2	.125
3	.125
	1

So $E(V) = 0(.5) + 1(.25) + 2(.125) + 3(.125) = .875$.

Therefore $\frac{E(U)}{E(V)} = \frac{.875}{.875} = 1$, so the ratio of male-to-female births would not change if every family in China followed the "one son" policy as there would be an equal number of male and female births in the long run. Had we changed the limit on the number of children in the family from 3 to some other number, both $E(U)$ and $E(V)$ would change but the ratio of $\frac{E(U)}{E(V)}$ would remain at 1. In Section 4.4, we will see that removing any restriction on the number of children in the family under the "one son" policy will still lead to a ratio $\frac{E(U)}{E(V)}$ of 1.

Next, let Y = total number of children in a family. Find the distribution of Y and then $E(Y)$.

$$P(Y = 1) \quad = \quad P(M) \qquad\qquad\qquad = \quad \frac{1}{2} = .5.$$

$$P(Y = 2) \quad = \quad P(FM) \qquad\qquad\qquad = \quad \frac{1}{2} \cdot \frac{1}{2} = .25.$$

$$P(Y = 3) \quad = \quad P(FFM) + P(FFF) \quad = \quad \frac{1}{2} \cdot \frac{1}{2} \cdot \frac{1}{2} + \frac{1}{2} \cdot \frac{1}{2} \cdot \frac{1}{2} = .25.$$

The distribution of Y is:

y	$P(Y = y)$
1	.5
2	.25
3	.25
	1

So $E(Y) = 1(.5) + 2(.25) + 3(.25) = 1.75$.

The long-run average number of children in a family is 1.75.

If we were to change the limit on the number of children in the family from three to some other number, then $E(Y)$ would also change. In Example 4.7 in Section 4.4 we will learn how to find $E(Y)$ when the condition limiting the number of children in the family to three is removed.

Example 4.3. A student with two grand goes to a gambling casino in the Bahamas. She plays a series of games where her probability of winning each game is $\frac{1}{3}$ and of losing the game is $\frac{2}{3}$. She bets one grand on each game until either she has accumulated four grand (enough to pay her tuition for the next year) or she goes broke.

Carry out 15 simulations of an experiment using the random number target to:

(i) Estimate her probability of accumulating four grand.

(ii) Estimate the expected number of games she plays in a series.

Solution: Select numbers from the random number target to correspond to the games in the series. If a number selected is 1 or 2 then she wins that game (with probability $\frac{1}{3}$). If a number selected is 3, 4, 5, or 6 then she loses that game (with probability $\frac{2}{3}$). For example, if the numbers selected in a series are 5, 2, 3, 6, then:

5 means she loses the first game reducing her total to 1 grand;

2 means she wins the second game increasing her total to 2 grand;

3 means she loses the third game reducing her total to 1 grand;

6 means she loses the fourth game reducing her total to 0 grand (she goes broke).

Thus, she goes broke in four games for this series. We now list the results for the 15 simulations.

Simulation Number	Numbers from Random Number Target	Series Outcome	Number of Games in the Series
1	5,2,3,6	goes broke	4
2	4,6	goes broke	2
3	3,1,5,1,2,1	accumulates 4 grand	6
4	6,4	goes broke	2
5	5,2,4,4	goes broke	4
6	3,6	goes broke	2
7	4,4	goes broke	2
8	2,5,1,3,1,2	accumulates 4 grand	6
9	6,1,5,1,2,4,3,2,5,2,2,1	accumulates 4 grand	12
10	3,2,1,1	accumulates 4 grand	4
11	1,3,1,6,4,2,2,4,4,4	goes broke	10
12	6,1,5,4	goes broke	4
13	2,5,3,1,4,6	goes broke	6
14	3,6	goes broke	2
15	5,4	goes broke	2

In the 15 simulations, she accumulates four grand four times. The estimate of the probability that she accumulates four grand is $\frac{4}{15} = .2667$. This is our answer to (i). While we don't have the analytical tools to solve for the true probability that she accumulates four grand, it turns out to be $\frac{1}{5} = .2$. Note that if she had bet two grand per game (instead of one grand), her chances of accumulating four grand would be $\frac{1}{3}$ which is greater than $\frac{1}{5}$. If the game is unfavorable to the player (probability of winning a game is less than .5), then the best strategy for this player is to bet the largest possible amount on each game.

The estimate of the expected number of games she plays until a decision is reached is

$$\frac{(4 + 2 + 6 + 2 + 4 + \cdots + 6 + 2 + 2)}{15} = \frac{68}{15} = 4.533.$$

This is our answer to (ii). Again, keep in mind that this is only an estimate.

Example 4.4. Suppose a machine has two working parts. Let X denote the number of years until part I malfunctions. Suppose the distribution of X is:

x	$P(X = x)$
1	$\frac{1}{3}$
2	$\frac{1}{3}$
3	$\frac{1}{3}$

Note that the expected number of years until part I malfunctions is

$$E(X) = 1\left(\frac{1}{3}\right) + 2\left(\frac{1}{3}\right) + 3\left(\frac{1}{3}\right) = 2.$$

Similarly, let W denote the number of years until part II malfunctions. If W has the same distribution as X, then $E(W) = 2$.

Suppose the machine will malfunction as soon as either one of the parts malfunctions. Let Y denote the number of years until the machine malfunctions.

If $x = 3$ and $w = 2$, then $y = 2$. This is because if part I malfunctions after 3 years and part II malfunctions after 2 years, the machine malfunctions as soon as either of the parts malfunctions which is after 2 years when part II malfunctions. Assume

$P(\text{part I malfunctions in } i \text{ years and part II malfunctions in } j \text{ years})$
$$= P(\text{part I malfunctions in } i \text{ years})$$
$$\cdot P(\text{part II malfunctions in } j \text{ years}).$$

The following table gives all possible values assumed by X and W. Then the value of y is the smaller of x and w. In all cases the probability is $\frac{1}{9}$ which is the product of the probabilities that parts I and II malfunction for each possible number of years.

x	w	y	Probability
1	1	1	$\frac{1}{3} \cdot \frac{1}{3} = \frac{1}{9}$
1	2	1	$\frac{1}{3} \cdot \frac{1}{3} = \frac{1}{9}$
1	3	1	$\frac{1}{3} \cdot \frac{1}{3} = \frac{1}{9}$
2	1	1	$\frac{1}{3} \cdot \frac{1}{3} = \frac{1}{9}$
2	2	2	$\frac{1}{3} \cdot \frac{1}{3} = \frac{1}{9}$
2	3	2	$\frac{1}{3} \cdot \frac{1}{3} = \frac{1}{9}$
3	1	1	$\frac{1}{3} \cdot \frac{1}{3} = \frac{1}{9}$
3	2	2	$\frac{1}{3} \cdot \frac{1}{3} = \frac{1}{9}$
3	3	3	$\frac{1}{3} \cdot \frac{1}{3} = \frac{1}{9}$

Summarizing the information in this table we see that the distribution of Y is:

y	$P(Y = y)$
1	$\frac{5}{9}$
2	$\frac{3}{9}$
3	$\frac{1}{9}$

The expected number of years until the machine malfunctions is

$$E(Y) = 1 \left(\frac{5}{9} \right) + 2 \left(\frac{3}{9} \right) + 3 \left(\frac{1}{9} \right) = 1.56.$$

One can show that if the machine has three working parts each with the same distribution until malfunction as X and the parts malfunction independently, the expected number of years until the machine malfunctions is 1.33. So more working parts for the machine means the expected number of years until the machine malfunctions gets smaller.

With modern technology, the expected length of time until a part malfunctions is longer than it was years ago. But counterbalancing that is the fact that machines are more complicated with more parts than previously, often causing the expected time until the machine malfunctions to be shorter.

4.2   Joint Distributions

In some problems we are concerned with the joint behavior of two random variables. Suppose that X and Y are random variables. Then $P(X = x, Y = y)$ is called the joint distribution of X and Y. The comma in $P(X = x, Y = y)$ denotes the word "and," so that $P(X = x, Y = y)$ generalizes $P(A \text{ and } B)$ in the same way that $P(X = x)$ generalizes $P(A)$.

Example 4.5. Three red blocks and two green blocks are arranged at random in a row. Let X denote the longest string of consecutive green blocks and Y denote the longest string of consecutive red blocks. We list all possible outcomes in the equally probable sample space along with the corresponding values of x and y for each outcome. Since there are 10 outcomes in S, each outcome has probability $\frac{1}{10}$.

Outcome	x	y
$RRRGG$	2	3
$RRGRG$	1	2
$RRGGR$	2	2
$RGGRR$	2	2
$RGRGR$	1	1
$RGRRG$	1	2
$GRRRG$	1	3
$GRRGR$	1	2
$GRGRR$	1	2
$GGRRR$	2	3

To find the joint distribution of X and Y, we need to find the probability that X and Y assume each possible pair of values.

$$P(X = 1, Y = 1) = P(RGRGR) = \tfrac{1}{10}.$$

$$
\begin{aligned}
P(X = 1, Y = 2) &= P(RRGRG \text{ or } RGRRG \text{ or } GRRGR \text{ or } GRGRR) \\
&= P(RRGRG) + P(RGRRG) \\
&\quad + P(GRRGR) + P(GRGRR) \\
&= \tfrac{1}{10} + \tfrac{1}{10} + \tfrac{1}{10} + \tfrac{1}{10} = \tfrac{4}{10}.
\end{aligned}
$$

$$P(X = 1, Y = 3) = P(GRRRG) = \tfrac{1}{10}.$$

$$P(X = 2, Y = 2) = P(RRGGR) + P(RGGRR) = \tfrac{1}{10} + \tfrac{1}{10} = \tfrac{2}{10}.$$

$$P(X = 2, Y = 3) = P(RRRGG) + P(GGRRR) = \tfrac{1}{10} + \tfrac{1}{10} = \tfrac{2}{10}.$$

Next assemble all this information in a two-way table giving the joint distribution of X and Y. The entry in the table corresponding to x and y is $P(X = x, Y = y)$.

$$
\begin{array}{cc|cc}
 & & \multicolumn{2}{c}{x} \\
 & & 1 & 2 \\
\hline
 & 1 & \frac{1}{10} & 0 \\
y & 2 & \frac{4}{10} & \frac{2}{10} \\
 & 3 & \frac{1}{10} & \frac{2}{10} \\
\end{array}
$$

This table gives us the joint distribution of X and Y. It tells us the probability that X and Y will assume all possible values. The sum of all the probabilities in the table must add up to 1 since all possibilities are represented.

Suppose we know the joint distribution of X and Y and would like to know the individual distribution of X. How could we find the distribution of X from the joint distribution of X and Y in Example 4.5?

$$
\begin{aligned}
P(X = 1) &= P(X = 1, Y = 1 \text{ or } X = 1, Y = 2 \text{ or } X = 1, Y = 3) \\
&= P(X = 1, Y = 1) + P(X = 1, Y = 2) \\
&\quad + P(X = 1, Y = 3) \\
&= \tfrac{1}{10} + \tfrac{4}{10} + \tfrac{1}{10} = \tfrac{6}{10}. \\
P(X = 2) &= P(X = 2, Y = 2) + P(X = 2, Y = 3) = \tfrac{2}{10} + \tfrac{2}{10} = \tfrac{4}{10}.
\end{aligned}
$$

So the column totals give us the distribution of X which is:

x	$P(X = x)$
1	$\frac{6}{10}$
2	$\frac{4}{10}$

Similarly, the row totals will give us the distribution of Y:

y	$P(Y = y)$
1	$\frac{1}{10}$
2	$\frac{6}{10}$
3	$\frac{3}{10}$

The individual distributions of one random variable are sometimes called marginal distributions because they are the totals found in the margins of the joint distribution table.

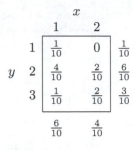

★ 4.3   Conditional Distributions

$P(Y = y|X = x)$ is called the conditional distribution of Y given x. It generalizes $P(B|A)$. Suppose in Example 4.5, we wanted to find $P(Y = y|X = 1)$. That is, if we know that the longest string of consecutive green blocks is 1, what is the distribution for the longest string of consecutive red blocks?

Look at a specific value of y, say $y = 3$: Let $A =$ "$X = 1$," $B =$ "$Y = 3$."

$$
\begin{aligned}
P(Y = 3|X = 1) &= P(B|A) = P(A \text{ and } B)/P(A) \\
&= P(X = 1 \text{ and } Y = 3)/P(X = 1) \\
&= \frac{1}{10} \Big/ \frac{6}{10} = \frac{1}{6}.
\end{aligned}
$$

Looking back at the joint distribution, we see that $P(Y = 3|X = 1)$ is simply the table entry at $X = 1, Y = 3$ divided by the marginal total for $X = 1$.

In addition, $P(Y = 2|X = 1) = \frac{4}{10}/\frac{6}{10} = \frac{4}{6}$ and $P(Y = 1|X = 1) = \frac{1}{10}/\frac{6}{10} = \frac{1}{6}$.

We can put this information in table form:

y	$P(Y = y|X = 1)$
1	$\frac{1}{6}$
2	$\frac{4}{6}$
3	$\frac{1}{6}$
	1

To complete the conditional distribution of Y given x for Example 4.5, we must also include $P(Y = y | X = 2)$:

y	$P(Y = y \mid X = 1)$
1	$\frac{1}{6}$
2	$\frac{4}{6}$
3	$\frac{1}{6}$
	1

y	$P(Y = y \mid X = 2)$
2	$\frac{2}{4}$
3	$\frac{2}{4}$
	1

Since a conditional distribution is a distribution, the probabilities must total 1.

Example 4.6. A fair coin is tossed three times. Let X denote the number of heads in the first two tosses and Y denote the number of heads in the last two tosses.

Find:

(i) the joint distribution of X and Y;

(ii) the marginal distribution of X;

(iii) the conditional distribution of Y given X.

Solution: We begin by listing all possible outcomes in the equally probable sample space along with the corresponding values of x and y.

Outcome	x	y
HHH	2	2
HHT	2	1
HTH	1	1
HTT	1	0
THH	1	2
THT	1	1
TTH	0	1
TTT	0	0

Each of the eight equally likely outcomes has probability 1/8.

(i) This information is compiled to give the joint distribution of X and Y:

		x		
		0	1	2
	0	$\frac{1}{8}$	$\frac{1}{8}$	0
y	1	$\frac{1}{8}$	$\frac{2}{8}$	$\frac{1}{8}$
	2	0	$\frac{1}{8}$	$\frac{1}{8}$

(ii) From (i), the marginal distribution of X is given by the column totals:

x	$P(X = x)$
0	$\frac{2}{8}$
1	$\frac{4}{8}$
2	$\frac{2}{8}$

(iii) The conditional distribution of Y given x is determined by dividing the joint probabilities in (i) by the marginals in (ii):

y	$P(Y = y \mid X = 0)$
0	$\frac{1}{2}$
1	$\frac{1}{2}$

y	$P(Y = y \mid X = 1)$
0	$\frac{1}{4}$
1	$\frac{1}{2}$
2	$\frac{1}{4}$

y	$P(Y = y \mid X = 2)$
1	$\frac{1}{2}$
2	$\frac{1}{2}$

★ 4.4 Conditional Expectation

Just as we calculated an expected value for a distribution in Section 4.1, we can calculate an expected value for a conditional distribution. The conditional expectation of Y given x, denoted by $E(Y|x)$, is found by summing all possible products of y times $P(Y = y|X = x)$. This will give us the center of gravity of the conditional distribution of Y for each possible value of x. For instance, in Example 4.5, we see by looking at the conditional distributions

y	$P(Y = y \mid X = 1)$
1	$\frac{1}{6}$
2	$\frac{4}{6}$
3	$\frac{1}{6}$

y	$P(Y = y \mid X = 2)$
2	$\frac{2}{4}$
3	$\frac{2}{4}$

that $E(Y|1) = 1\left(\frac{1}{6}\right)+2\left(\frac{4}{6}\right)+3\left(\frac{1}{6}\right) = 2$ and $E(Y|2) = 2\left(\frac{2}{4}\right)+3\left(\frac{2}{4}\right) = 2.5$.

A result in probability states that $E(Y)$ may be calculated by taking a weighted average of the $E(Y|x)$ values, each value weighted by the probability of the event on which it is conditioned. To verify that this result is satisfied in Example 4.5, we calculate $E(Y)$ first by using this method and then by direct calculation.

$$E(Y) = E(Y|1)P(X = 1)+E(Y|2)P(X = 2) = 2\left(\frac{6}{10}\right)+2.5\left(\frac{4}{10}\right) = 2.2.$$

From the distribution of Y,

y	$P(Y = y)$
1	$\frac{1}{10}$
2	$\frac{6}{10}$
3	$\frac{3}{10}$

we determine that $E(Y) = 1\left(\frac{1}{10}\right) + 2\left(\frac{6}{10}\right) + 3\left(\frac{3}{10}\right) = 2.2$.

We see that the result holds for this example.

The following example illustrates how this result can be used to determine an expected value when Y can assume an infinite number of values.

Example 4.7. We refer back to Example 4.2. Specifically consider the "one son" policy but remove the condition limiting the number of children in the family. Let Y = total number of children in a family. Find $E(Y)$.

Outcome	y	$P(Y = y)$
M	1	$\frac{1}{2}$
FM	2	$\frac{1}{4}$
FFM	3	$\frac{1}{8}$
$FFFM$	4	$\frac{1}{16}$
.	.	.
.	.	.
.	.	.

The problem here is that there are an infinite number of values for y, one for each positive integer. We could get an approximate value for $E(Y)$ by including a sufficiently large number of terms.

But to get an exact answer for $E(Y)$ we introduce another random variable. Let X denote the number of males on the first birth. That is, $X = 0$ if the first child is F and $X = 1$ if the first child is M. The only possible values that X can assume are 0 and 1. Random variables that only assume the values 1 and 0 occur frequently in probability. They represent situations where an event either does or does not occur. According to our result,

$$E(Y) = E(Y|0)P(X = 0) + E(Y|1)P(X = 1).$$

If $X = 1$, then we have a male on the first birth and so the only possible value Y can assume is 1. Thus, $E(Y|1) = 1$. If $X = 0$, we have a female on the first birth. So the various possible outcomes until we get a male are FM, FFM, $FFFM$, etc. Since each of these outcomes has one more child than M, FM, FFM, where the distribution has expected value $E(Y)$, we conclude that $E(Y|0) = E(Y) + 1$.

Substituting into the equation

$$E(Y) = E(Y|0)P(X = 0) + E(Y|1)P(X = 1)$$

gives

$$E(Y) = [E(Y) + 1]P(X = 0) + 1P(X = 1)$$

or

$$E(Y) = [E(Y) + 1]\frac{1}{2} + 1\left(\frac{1}{2}\right).$$

Solving for $E(Y)$ gives $E(Y) = \frac{1}{2}E(Y) + 1$.

$$\frac{1}{2}E(Y) = 1;$$

$$E(Y) = 2.$$

So if families have children until they have a son, the average number of children in a family will approach two in the long run.

4.5 Problems

1. Two red blocks and two green blocks are arranged at random in a row. Let $X=$ number of green blocks between the two red blocks.

   (i) Find the distribution of X by first listing all outcomes in S.
   (ii) Find $E(X)$.

2. Give an example of a random variable in each of the following problems:

   (i) Problem 6 in Chapter 1.
   (ii) Problem 7 in Chapter 1.

3. Four candidates, two men and two women, are interviewing for a job. The candidates are to be divided randomly between two personnel officers, who will interview two candidates apiece. Let X denote the number of women assigned to the first personnel officer. Find the distribution of X.

4. An integer is chosen at random from among the integers 1,2,3,4,5,6. Let X be the number of different positive integers that divide the chosen integer without remainder (For example, if the integer 6 is chosen, then $x = 4$ since 6 can be divided without remainder by 1,2,3, and 6). Find the distribution of X.

5. Refer to Example 4.1. Find $E(X)$ if the Braves and the Redsox are evenly matched.

6. Refer to Example 4.1. Let $P(B) = p$.

   (i) For what value(s) of p is $E(X)$ largest?
   (ii) For what value(s) of p is $E(X)$ smallest?

7. A group of 2 men and 3 women are lined up at random. Find the expected number of women that have a man on at least one side of them.

8. Refer back to Problem 6 in Chapter 1. List outcomes in S to solve this problem.

   (i) Three letters written to Smith, Jones, and Clark are placed at random in three envelopes addressed to Smith, Jones, and Clark with one letter in each envelope. Find the expected number of letters that go into the correct envelope.

   (ii) Now, suppose there are only two people involved, call them Smith and Jones. Two letters written to Smith and Jones are placed at random in two envelopes addressed to Smith, Jones with one letter in each envelope. Find the expected number of letters that go into the correct envelope.

9. A biased coin, where $P(\text{head}) = \frac{4}{5}$, is to be flipped until two consecutive heads appear. Estimate the expected number of tosses it will take for that to happen by carrying out 10 simulations using the random number target.

10. Jennifer has tickets to the fourth and fifth games of the World Series of Women's Lacrosse. In this World Series, the first team to win four games wins the series. If the two teams playing in the series are evenly matched, find the expected number of games Jennifer will see.

11. Refer back to Problem 14 in Chapter 1. Find the expected rank of the runner-up.

12. A total of three buses carrying 75 Bucknell football players arrives at Ball State for the big game. The buses carry, respectively, 5, 30, 40 players. One of the players is randomly selected. Let X denote the number of players that were on the bus carrying this randomly selected player. One of the three bus drivers is also randomly selected. Let Y denote the number of players on her bus. Compute $E(X)$ and $E(Y)$.

13. Assume that a basketball player has probability .7 of making a free throw and that each free throw is independent of previous throws. Find the expected number of points the player will make if:

(i) The player is awarded two free throws.

(ii) A one-and-one free throw situation exists. That is, if the first shot is made the player gets a second shot; if the first shot is missed there is no second shot.

14. An academic building on the University of Pittsburgh campus has seven floors above the ground level. There are two elevators on the ground level where students enter the building. Elevator #1 can stop at floors numbered 2, 4, 6 and elevator #2 can stop at floors numbered 1, 3, 5, 7. Suppose two students enter elevator #1 on the ground level and exit the elevator independently and equally likely at each of the three stops. Determine the expected number of stops until elevator #1 is empty.

15. Suppose a machine has two working parts and the distribution of the number of years until each part malfunctions (X) is:

x	$P(X = x)$
1	$\frac{1}{3}$
2	$\frac{2}{3}$

The machine will malfunction as soon as either of the parts malfunctions. Find the expected number of years until the machine malfunctions and compare it with $E(X)$.

16. Let X denote the number of machine failures per day in a certain plant. Suppose X has distribution:

x	$P(X = x)$
0	$\frac{1}{3}$
1	$\frac{1}{6}$
2	$\frac{1}{6}$
3	$\frac{1}{6}$
4	$\frac{1}{6}$

Present maintenance facilities at the plant can repair two machines per day. Failures in excess of two are repaired by a contractor.

(i) Find the expected number of machine failures per day at the plant.

(ii) Find the expected number of machines repaired *at the plant* per day.

17. In Example 4.4, find:

   (i) The joint distribution of X and Y.

   (ii) The marginal distribution of X from (i).

   (iii) The marginal distribution of Y from (i).

18. Let X be the number of birthmonths among three randomly selected persons. Assume all 12 birthmonths are equally likely. Find $E(X)$.

19. Two red blocks and two green blocks are arranged at random in a row.

   Let X = the number of green blocks between the two red blocks.

   Let Y = the number of red blocks between the two green blocks.

   Find the joint distribution of X and Y.

*20. A fair coin is tossed three times. Let X be the number of heads on the first two tosses. Let Y be the number of heads on all three tosses.

   (i) Find the joint distribution of X and Y.

   (ii) Find $P(X = 2|Y = 2)$.

*21. Two balls are distributed at random into two boxes. Let X be the number of empty boxes, and let Y be the number of balls in the first box. Find:

   (i) the joint distribution of X and Y,

   (ii) the marginal distribution of X,

   (iii) the conditional distribution of Y given $X = 1$,

   (iv) $P(X \leq Y)$.

22. Food Lion has two express check out lines. Let X denote the number of customers in the first line and Y the number in the second line. During non-rush hours, the joint distribution of X and Y is given by the following table:

$$x$$

		0	1	2
	0	.1	.04	.01
y	1	.04	.2	.15
	2	.01	.15	.3

(i) Find the marginal distribution of X.

(ii) Would you expect $P(X = x, Y = y)$ to be symmetric? Why?

(iii) Would you expect $P(|X - Y| > 1)$ to be small? Why?

23. A number is selected at random from the set $\{1, 2\}$. This number is denoted by X. Then x balls are selected with replacement from a box containing three red and two white balls. Let Y denote the number of white balls drawn. Find the joint distribution of X and Y.

24. York County, Virginia, mails out the real estate tax bill and car tax bill to residents at the same time. Each bill is sent out with an identically addressed return envelope. So each resident has a choice of sending back each payment in one envelope (two envelopes in total) or both payments in the same envelope (one envelope in total).

Suppose $p = P(\text{an envelope gets lost in the mail})$.

Let X denote the number of checks that arrive at the county office for each resident.

In terms of p:

(i) Find the distribution of X and $E(X)$ if the resident uses one envelope.

(ii) Find the distribution of X and $E(X)$ if the resident uses two envelopes.

*25. A man tosses a fair die until a five appears. Find the expected number of tosses needed.

*26. A man plays Russian roulette using one bullet. After each trigger pull, he spins the cylinder. Find the expected number of trigger pulls until he kills himself. On which trigger pull is he most likely to kill himself?

27. Estimate $E(Y)$ in Example 4.2 by carrying out 10 simulations using the random number target. Assume there is no limit on the number of children a family may have under the "one son" policy.

28. Refer to Example 4.2. In the "one son" policy in China, suppose we specify that there will be at most two children in the family. Find $E(Y)$ and the ratio $\frac{E(U)}{E(V)}$.

*29. In Example 4.7, show that $E(V) = 1$ when there is no limit on the number of children a family may have under the "one son" policy. Also show that $E(U) = 1$, so that the ratio $\frac{E(U)}{E(V)} = 1$.

5

Sampling without Replacement

5.1 Counting Formula

In this section we develop the counting formula needed for Chapters 5 and 6.

Definition: *Factorial n* is denoted by $n!$ and given by

$$n! = n(n-1)(n-2)\cdots 3\cdot 2\cdot 1$$

for n a positive integer. So $n!$ is the product of the first n positive integers where n is a positive integer. We define $0! = 1$.

There are various ways of expressing a factorial. For example,

$$
\begin{aligned}
5! &= 5\cdot 4\cdot 3\cdot 2\cdot 1, \\
5! &= 5(4!), \\
5! &= 5\cdot 4(3!), \\
5! &= 5\cdot 4\cdot 3(2!).
\end{aligned}
$$

We can use this information in evaluating ratios of factorials. For example,

$$\frac{5!}{3!2!} = \frac{5\cdot 4(3!)}{3!2!} = \frac{5\cdot 4}{2\cdot 1} = 10.$$

Important Note: For some of the material that follows, the computations are tedious. When this turns out to be the case, our objective will be to get an expression for the answer. Accordingly, the statement, "Find an expression for" means the answer should be given in terms of sums, products, quotients, factorials, powers, etc.

Example 5.1. A box contains three chips labelled A and two chips labelled B. All five chips are selected without replacement from the box. Think of every possible ordered selection as a word. How many words are there? We answer this question with the help of the multiplication rule.

Look at the word $AABAB$. The probability of drawing an A as the first letter is $\frac{3}{5}$. The conditional probability that the second letter is A given that the first letter is A is $\frac{2}{4}$. If the first two letters are A, there is one chip labelled A and two chips labelled B remaining in the box. So the probability of getting a B given the first two chips are A's is $\frac{2}{3}$. Continuing this argument, by the multiplication rule the probability of getting $AABAB$ is $\frac{3}{5} \cdot \frac{2}{4} \cdot \frac{2}{3} \cdot \frac{1}{2} \cdot \frac{1}{1}$. We now begin making a list of words with $3A$'s, $2B$'s, and their corresponding probabilities.

Word	P(Word)		
$AABAB$	$\frac{3}{5} \cdot \frac{2}{4} \cdot \frac{2}{3} \cdot \frac{1}{2} \cdot \frac{1}{1}$	$= \frac{3!2!}{5!}$	$= \frac{1}{10}$
$AAABB$	$\frac{3}{5} \cdot \frac{2}{4} \cdot \frac{1}{3} \cdot \frac{2}{2} \cdot \frac{1}{1}$	$= \frac{3!2!}{5!}$	$= \frac{1}{10}$
$ABBAA$	$\frac{3}{5} \cdot \frac{2}{4} \cdot \frac{1}{3} \cdot \frac{2}{2} \cdot \frac{1}{1}$	$= \frac{3!2!}{5!}$	$= \frac{1}{10}$
.	.		
.	.		
.	.		

We see that each word has probability $\frac{3!2!}{5!} = \frac{1}{10}$. Since exactly one of these words must appear,

$$P(AABAB) \;+\; P(AAABB) \;+\; P(ABBAA) \;+\cdots \;=\; 1;$$

$$\tfrac{1}{10} \quad + \quad \tfrac{1}{10} \quad + \quad \tfrac{1}{10} \quad +\cdots \;=\; 1.$$

We must add $\frac{1}{10}$ a total of ten times to get 1. Thus, the number of words with $3A's$, $2B's$ is

$$10 = \frac{5!}{3!2!} = \frac{(\text{total number of letters})!}{(\text{number of }A\text{'s})!(\text{number of }B\text{'s})!}.$$

In general, the number of words we can form with n_1 A's and n_2 B's is

$$\frac{(n_1 + n_2)!}{n_1! n_2!}.$$

We can extend this result for any number of letters. For three different letters, the number of words with n_1 A's, n_2 B's, and n_3 C's is

$$\frac{(n_1 + n_2 + n_3)!}{n_1!n_2!n_3!}.$$

This gives us the counting formula we need for the material that follows.

For the remainder of this chapter and also in Chapter 6, we will study some widely used probability models for applications which frequently arise in practice. The development of these models is based on the addition and multiplication rules.

5.2 Probabilities for Sampling without Replacement

Example 5.2. A box contains eight red and six white chips. Four chips are drawn at random and without replacement from the box. Let X denote the number of red chips drawn. Find an expression for the probabilities of the following events:

(i) three red chips followed by one white chip;

(ii) one red chip followed by one white chip and then two red chips;

(iii) $X = 3$;

(iv) $X \geq 3$.

Solution:

(i) P(3 red chips followed by 1 white chip) $= P(RRRW) = \frac{8}{14} \cdot \frac{7}{13} \cdot \frac{6}{12} \cdot \frac{6}{11}$ using the multiplication rule.

(ii) P(1 red chip followed by 1 white chip and then 2 red chips) $= P(RWRR) = \frac{8}{14} \cdot \frac{6}{13} \cdot \frac{7}{12} \cdot \frac{6}{11}$. Note that the answers to (i) and (ii) are the same because both answers have the same factors in the numerators and in the denominators. This will always be the case so long as we draw one white and three red chips.

(iii) $P(X = 3) = P$(3 red chips and 1 white chip in any order)
$$= P(RRRW) + P(RRWR) + P(RWRR) + P(WRRR).$$

The number of terms in the sum is equal to the number of words with $3R$'s, $1W$ because each term corresponds to a different word. But, from Section 5.1 we know that the number of words with $3R$'s, $1W$ is $\frac{(3+1)!}{3!1!} = \frac{4!}{3!1!} = \frac{4(3!)}{3!1!} = 4$. As indicated in (ii), each of the four probabilities are equal. Thus,

$$
\begin{aligned}
P(X = 3) &= P(RRRW) + P(RRWR) \\
&\quad + P(RWRR) + P(WRRR) \\
&= 4P(RRRW) \\
&= \frac{4!}{3!1!} P(RRRW) \\
&= \frac{4!}{3!1!} \left(\frac{8}{14} \cdot \frac{7}{13} \cdot \frac{6}{12} \cdot \frac{6}{11} \right).
\end{aligned}
$$

(iv) $P(X \geq 3) = P(X = 3) + P(X = 4)$ by the addition rule.

$$
P(X = 4) = \frac{4!}{4!0!} \left(\frac{8}{14} \cdot \frac{7}{13} \cdot \frac{6}{12} \cdot \frac{5}{11} \right).
$$

The quantity in parentheses is the probability of four reds and the coefficient of $\frac{4!}{4!0!} = 1$ is the number of words with $4R$'s, $0W$'s. Thus

$$
P(X \geq 3) = \frac{4!}{3!1!} \left(\frac{8}{14} \cdot \frac{7}{13} \cdot \frac{6}{12} \cdot \frac{6}{11} \right) + \frac{4!}{4!0!} \left(\frac{8}{14} \cdot \frac{7}{13} \cdot \frac{6}{12} \cdot \frac{5}{11} \right).
$$

Note that in an example such as this where we sample without replacement, the draws are not independent. So events such as "red chip on first draw" and "red chip on second draw" are not independent events.

Example 5.3. A box contains 10 chips numbered $1, 2, 3, \ldots, 10$. Five chips are selected at random without replacement. Find an expression for the probability that the:

(i) smallest number drawn is 4;

(ii) median number drawn is 4.

Solution: For both parts of the problem the number 4 must be drawn. In addition a certain number of chips smaller than 4 and a certain number larger than 4 must be drawn. So the numbers in the box fall into three categories: G(greater than 4), L(less than 4), E(equal to 4). Relabel each of the 10 chip numbers with letters as follows:

$$\begin{array}{cccccccccc} 1 & 2 & 3 & 4 & 5 & 6 & 7 & 8 & 9 & 10 \\ L & L & L & E & G & G & G & G & G & G \end{array}$$

Now we are ready to answer the question.

(i) P(smallest number drawn is 4)

$$= P(1E, 4G\text{'s})$$

$$= \frac{5!}{1!4!} P(EGGGG)$$

$$= \frac{5!}{1!4!} \left(\frac{1}{10} \cdot \frac{6}{9} \cdot \frac{5}{8} \cdot \frac{4}{7} \cdot \frac{3}{6} \right).$$

If you don't relabel the chip numbers with letters, it is easy to forget that there are $\frac{5!}{1!4!} = 5$ words with $1E, 4G$'s and give $\frac{1}{10} \cdot \frac{6}{9} \cdot \frac{5}{8} \cdot \frac{4}{7} \cdot \frac{3}{6}$ as the final answer.

(ii) If the median number drawn is 4, there must be two numbers less than the median and two numbers greater than the median in the sample. There are $\frac{(2+1+2)!}{2!1!2!} = \frac{5!}{2!1!2!}$ words with $2L$'s, $1E$, $2G$'s.

$$P(\text{median number is 4})$$

$$= P(2L\text{'s}, 1E, 2G\text{'s})$$

$$= \frac{5!}{2!1!2!} P(LLEGG)$$

$$= \frac{5!}{2!1!2!} \left(\frac{3}{10} \cdot \frac{2}{9} \cdot \frac{1}{8} \cdot \frac{6}{7} \cdot \frac{5}{6} \right).$$

Example 5.4. In a close election in a small town, 637 people voted for candidate A compared to 630 people for candidate B, a margin of seven votes. An investigation found that ten people who voted in the election should not have (we don't know who they voted for). This is more than the margin of victory. What is the probability that the random removal of ten votes would reverse the election results?

Solution: First we look for the split of improperly cast votes which would change the result if removed from the tally. If we removed all ten votes from A then the election result would be reversed with B winning 630 to 627. If we removed nine votes from A and one vote from B then the election result would be reversed with B winning 629 to 628. However, if we removed eight votes from A and two votes from B then the election result would not be reversed and A would win 629 to 628. If we removed fewer than eight votes from A, then A would win.

Imagine a box with 637 chips labelled A(corresponding to A's votes) and 630 chips labelled B(corresponding to B's votes). Remove ten votes at random without replacement from the box. If we selected at least nine A's, the election results would be reversed.

P(random removal of ten votes will reverse election results)

$$= \quad P(9A's, 1B) + P(10A's)$$

$$= \quad \frac{10!}{9!1!}P(AAAAAAAAAB) + \frac{10!}{10!0!}P(AAAAAAAAAA)$$

$$= \quad \frac{10!}{9!1!}\left(\frac{637}{1267}\right)\left(\frac{636}{1266}\right)\cdots\left(\frac{629}{1259}\right)\left(\frac{630}{1258}\right)$$

$$\quad + \frac{10!}{10!0!}\left(\frac{637}{1267}\right)\left(\frac{636}{1266}\right)\cdots\left(\frac{628}{1258}\right)$$

$$= \quad .011.$$

The probability that the removal of ten randomly chosen votes will reverse the election results is .011. It is highly unlikely that the election result would have been different if the ten people had not voted.

Example 5.5. Each of two precincts has 100 voters. The number of Democrats in precinct I is 20 and in precinct II is 80. Assume the other voters are Republicans. For each of the following methods of selecting two voters, find the distribution of X, the number of Democrats selected.

 (i) Select one voter at random from each precinct.

 (ii) Select two voters at random without replacement from the combined group of 200 voters.

 (iii) Select one of the two precincts at random, then select two voters at random without replacement from the selected precinct.

Solution: Call the voters D for Democrat and R for Republican, and the precincts I and II.

(i)

$$P(X = 0) = P(R \text{ from } I \text{ and } R \text{ from } II)$$

$$= \frac{80}{100} \cdot \frac{20}{100}$$

$$= .16.$$

$$P(X = 1) = P(R \text{ from } I \text{ and } D \text{ from } II)$$

$$+ P(D \text{ from } I \text{ and } R \text{ from } II)$$

$$= \frac{80}{100} \cdot \frac{80}{100} + \frac{20}{100} \cdot \frac{20}{100}$$

$$= .68.$$

$$P(X = 2) = \frac{20}{100} \cdot \frac{80}{100}$$

$$= .16.$$

So the distribution of X is:

x	$P(X = x)$
0	.16
1	.68
2	.16

(ii)

$$P(X = 0) = P(RR)$$

$$= \left(\frac{2!}{0!2!}\right)\left(\frac{100}{200}\right)\left(\frac{99}{199}\right)$$

$$= .24874.$$

$$P(X = 1) = P(1D \text{ and } 1R)$$

$$= \left(\frac{2!}{1!1!}\right)\left(\frac{100}{200}\right)\left(\frac{100}{199}\right)$$

$$= .50252.$$

The reason we use a ratio of factorials term in (ii) but not in (i) is because in (ii) we are sampling from the combined group, as opposed to two different groups, so $P(DR) = P(RD)$.

$$P(X = 2) = P(DD)$$

$$= \left(\frac{2!}{2!0!}\right)\left(\frac{100}{200}\right)\left(\frac{99}{199}\right)$$

$$= .24874.$$

So the distribution of X is:

x	$P(X = x)$
0	.24874
1	.50252
2	.24874

(iii)

$$P(X = 0) = P(I \text{ and } RR) + P(II \text{ and } RR)$$

$$= P(I)P(RR|I) + P(II)P(RR|II)$$

$$= \frac{1}{2}\left(\frac{2!}{0!2!} \cdot \frac{80}{100} \cdot \frac{79}{99}\right) + \frac{1}{2}\left(\frac{2!}{0!2!} \cdot \frac{20}{100} \cdot \frac{19}{99}\right)$$

$$= .33838.$$

$$P(X = 1) = P(I \text{ and } 1D, 1R) + P(II \text{ and } 1D, 1R)$$

$$= \frac{1}{2}\left(\frac{2!}{1!1!} \cdot \frac{20}{100} \cdot \frac{80}{99}\right) + \frac{1}{2}\left(\frac{2!}{1!1!} \cdot \frac{80}{100} \cdot \frac{20}{99}\right)$$

$$= .32324.$$

$$P(X = 2) = P(I \text{ and } DD) + P(II \text{ and } DD)$$

$$= \frac{1}{2}\left(\frac{2!}{2!0!} \cdot \frac{20}{100} \cdot \frac{19}{99}\right) + \frac{1}{2}\left(\frac{2!}{2!0!} \cdot \frac{80}{100} \cdot \frac{79}{99}\right)$$

$$= .33838.$$

So the distribution of X is:

x	$P(X = x)$
0	.33838
1	.32324
2	.33838

An interesting observation is that for all three sampling plans $E(X) = 1$, yet the distributions are different. The distribution is more concentrated at $X = 1$ for plan (i) than for plan (ii) and is more concentrated at $X = 1$ for plan (ii) than for plan (iii).

The following is another version of the Capture-Recapture problem which was presented in Example 3.8.

Example 5.6. (Capture-Recapture Problem (Non-Bayes' Version).) We want to estimate the number of fish in a pond. We catch three fish at random (without replacement), tag them, and then release them back into the pond. A day later we catch five fish and we observe that two of the five are tagged. Estimate the number of fish in the pond as the number which will maximize the probability of what we actually observed.

Solution: Let T denote "tagged," U "untagged," and let X be a random variable representing the number of tagged fish drawn. We solve the problem by trial and error.

Suppose there are five fish in the pond:

Since three fish were tagged, the contents of the pond are $3T, 2U$. Next we catch five fish without replacement. $P(X = 2) = 0$ since it is not possible to get two tagged fish and three untagged fish from this pond.

Suppose there are six fish in the pond:

Since three fish were tagged, the contents of the pond are $3T, 3U$. Next we catch five fish without replacement.

$$P(X = 2) = P(2T\text{'s}, 3U\text{'s}) = \frac{5!}{2!3!} P(TTUUU)$$

$$= \frac{5!}{2!3!} \left(\frac{3}{6} \cdot \frac{2}{5} \cdot \frac{3}{4} \cdot \frac{2}{3} \cdot \frac{1}{2} \right) = .500.$$

Suppose there are seven fish in the pond:

Since three fish were tagged, the contents of the pond are $3T$, $4U$. Next we catch five fish without replacement.

$$P(X = 2) = P(2T\text{'s, } 3U\text{'s}) \quad = \quad \frac{5!}{2!3!} P(TTUUU)$$

$$= \quad \frac{5!}{2!3!} \left(\frac{3}{7} \cdot \frac{2}{6} \cdot \frac{4}{5} \cdot \frac{3}{4} \cdot \frac{2}{3} \right) = .571.$$

Suppose there are eight fish in the pond:

Since three fish were tagged, the contents of the pond are $3T$, $5U$. Next we catch five fish without replacement.

$$P(X = 2) = P(2T\text{'s, } 3U\text{'s}) \quad = \quad \frac{5!}{2!3!} P(TTUUU)$$

$$= \quad \frac{5!}{2!3!} \left(\frac{3}{8} \cdot \frac{2}{7} \cdot \frac{5}{6} \cdot \frac{4}{5} \cdot \frac{3}{4} \right) = .536.$$

Suppose there are nine fish in the pond:

Since three fish were tagged, the contents of the pond are $3T$, $6U$. Next we catch five fish without replacement.

$$P(X = 2) = P(2T\text{'s, } 3U\text{'s}) \quad = \quad \frac{5!}{2!3!} P(TTUUU)$$

$$= \quad \frac{5!}{2!3!} \left(\frac{3}{8} \cdot \frac{2}{8} \cdot \frac{6}{7} \cdot \frac{5}{6} \cdot \frac{4}{5} \right) = .476.$$

We could continue on in this way but we would find that as the number of fish in the pond increases beyond nine, $P(X = 2)$ decreases. Which pond is most likely to have produced the sample which contains two tagged fish? By inspecting $P(X = 2)$ values, we see that the pond with seven fish is most likely to have produced the data. We estimate the number of fish in the pond to be seven. The pond with seven fish is the one which is "most consistent" with the data.

Example 5.7. In 1920, there were eight teams in the National League in Major League Baseball. Brooklyn won the pennant that year. Here are the final standings:

1. Brooklyn (B)

2. New York (N)

3. Cincinnati (C)

4. Pittsburgh

5. Chicago

6. St. Louis

7. Boston

8. Philadelphia

   In 1920, the top three teams in the league were money winners in that they received a portion of the World Series ticket receipts. If a sportswriter were predicting the money winners in the National League before the start of the 1920 season, what is the probability that he would correctly predict all three money winners if the eight teams were evenly matched?

Solution: Of the eight teams, three are money winners (W) and the other five are losers (L). So we can think of the sportswriter as selecting three letters at random from 3 W's and 5 L's.

P(sportswriter correctly predicts all three money winners)

$$= P(3\,W\text{'s}, 0\,L\text{'s}) = \frac{3!}{3!0!}\left[\frac{3}{8}\cdot\frac{2}{7}\cdot\frac{1}{6}\right] = \frac{1}{56}.$$

   Consider a second approach to this problem:
   If the sportswriter were predicting the teams that finished first, second, third in that order, then

P(sportswriter predicts that B is first, N is second, C is third)

$$= P(B, N, C) = \frac{1}{8}\cdot\frac{1}{7}\cdot\frac{1}{6}.$$

The first entry in the ordered triple, (B, N, C), is the team predicted to finish first, the second entry is the team predicted to finish second, and the third entry is the team predicted to finish third.

So,

$$P(\text{sportswriter correctly predicts all three money winners})$$

$$= \quad P(B, N, C) + P(N, C, B) + P(B, C, N) + \cdots$$

$$= \quad \frac{3!}{1!1!1!} \left[\frac{1}{8} \cdot \frac{1}{7} \cdot \frac{1}{6} \right] = \frac{1}{56}.$$

The first approach is simpler than the second approach because the three money winners are placed in one category (W).

Example 5.8. (Detecting a Drug User.)   There are many segments of the population where people are job tested on a regular basis for drug use. Jobs involving security, athletics, and the military are examples. We will explore a model for detecting drug use by random urinalysis testing and we will determine the probability of detecting drug use.

Suppose we take a time period of 30 days and during that period the drug user takes drugs on three different days selected at random from the 30 days. Assume that the probability of testing positive $(+)$ depends on the time since the most recent drug use. So the probability of testing $+$ does not depend on the pattern of drug use prior to testing but rather on the day that the user last takes drugs prior to testing. Assume the user takes drugs at a time of day which follows the scheduled time of day for testing. For this example, assume that the drug wears off over a two-day period.

We begin by looking at the case where there is a drug test one day per 30 days. Suppose that day #9 is the test day.

Since the drug wears off after two days, by the addition and multiplication rules,

$P(\text{user tests } +)$

$= \quad P(\text{user tests } + \text{ on day #9})$

$= \quad P(\text{user last takes drugs on day #8 and user tests } + \text{ on day #9})$

$\quad + P(\text{user last takes drugs on day #7 and user tests } + \text{ on day #9})$

$= \quad P(\text{user last takes drugs on day #8})$

$\quad \cdot \ P(\text{user tests } + \text{ on day #9}|\text{user last takes drugs on day #8})$

$\quad + P(\text{user last takes drugs on day #7})$

$\quad \cdot \ P(\text{user tests } + \text{ on day #9}|\text{user last takes drugs on day #7}).$

We must determine the four probabilities on the right side of the last equation in order to find $P(\text{user tests } +)$.

The wear-off probabilities for cocaine using random urinalysis drug testing were estimated (see reference [9] in the bibliography) using drug excretion rate kinetics as follows:

$$P(\text{user tests} + \text{ if tested one day after drug use}) = .938, \text{ and}$$
$$P(\text{user tests} + \text{ if tested two days after drug use}) = .651.$$

Since we assumed that day #9 is the test day,

$$P(\text{user tests} + \text{ on day } \#9 | \text{user last takes drugs on day } \#8) = .938;$$
$$P(\text{user tests} + \text{ on day } \#9 | \text{user last takes drugs on day } \#7) = .651.$$

Next we determine $P(\text{user last takes drugs on day } \#8)$. The user is selecting three days for drug use at random, without replacement, from 30 days. To satisfy the event "user last takes drugs on day #8," the user must take drugs on day #8 and on any two of the remaining 29 days. Label day #8 as A and the other 29 days as B.

$$P(\text{user last takes drugs on day } \#8) = P(1A, 2B\text{'s}) = \frac{3!}{2!1!}P(BBA)$$

$$= \frac{3!}{2!1!}\left(\frac{29}{30} \cdot \frac{28}{29} \cdot \frac{1}{28}\right) = .1.$$

Lastly we determine

$$P(\text{user last takes drugs on day } \#7)$$
$$= P(\text{user takes drugs on day } \#7 \text{ but not on day } \#8).$$

The user must take drugs on day #7 but not on day #8. Label day #8 as A, day #7 as B, and the other 28 days as C.

$$P(\text{user last takes drugs on day } \#7)$$

$$= P(\text{user takes drugs on day } \#7 \text{ but not on day } \#8)$$

$$= P(0A\text{'s}, 1B, 2C\text{'s})$$

$$= \frac{3!}{0!1!2!}P(BCC)$$

$$= \frac{3!}{0!1!2!}\left(\frac{1}{30} \cdot \frac{28}{29} \cdot \frac{27}{28}\right) = .0931.$$

Substituting,

P(user tests $+$)

= P(user last takes drugs on day #8)

· P(user tests $+$ on day #9|user last takes drugs on day #8)

+ P(user last takes drugs on day #7)

· P(user tests $+$ on day #9|user last takes drugs on day #7)

$$= \frac{3!}{2!1!}\left(\frac{29}{30}\cdot\frac{28}{29}\cdot\frac{1}{28}\right)(.938) + \frac{3!}{0!1!2!}\left(\frac{1}{30}\cdot\frac{28}{29}\cdot\frac{27}{28}\right)(.651)$$

$$= (.1)(.938) + (.0931)(.651) = .1544.$$

To this point, we have considered the case of one test day per 30 days. Let us consider a more general case. Suppose that 20% of the population are tested in 30 days (this is approximately the rate in the U.S. Navy). And suppose there are four test days in 30 days. This means that about $(.20)/4 = .05 = 1/20$ of the population are tested each test day.

P(user is detected on a test day)

= P(user is selected for testing on a test day and tests $+$)

= P(user is selected for testing on a test day)

· P(user tests $+$|user is selected for testing on a test day)

= $(.05)(.1544) = .0077$ by the multiplication rule.

P(user is not detected on a test day) $= 1 - .0077 = .9923$.

P(user is not detected on four test days)

= P(user is not detected in a 30-day period)

= $(.9923)^4 = .9696$ by the multiplication rule.

Since a 30-day period is approximately one month,

$$P(\text{user is detected in one month}) = 1 - .9696 = .0304.$$

Therefore using this model in a drug detection system, it can take several years before a drug user is detected.

5.3 Problems

1. (i) Use the formula to count the number of words with 2 A's and 2 B's.

   (ii) Verify your answer to (i) by listing all words with 2 A's and 2 B's.

2. A scrabble set consists of 54 consonants and 44 vowels. Find an expression for (do not evaluate) the probability that your initial draw of seven letters will consist of five consonants and two vowels.

3. Twenty-six white nurses and nine black nurses took an examination. All white nurses and four black nurses passed. *Dendy v. Washington Hospital Ctr.*, 431 F. Supp. 873 (D.D.C. 1977). Is it likely that such a large difference in pass rates is due to chance? Hint: Suppose five nurses are selected at random to fail the exam. What is the probability that all five selected are black nurses? Is it possible to solve this problem by enumeration using Chapter 1 techniques? If so, why don't we solve the problem using that method?

4. There are 97 men and three women in an organization. A committee of five people is chosen at random, and one of these five is randomly designated as chairperson. Find an expression for (do not evaluate) the probability that the committee

   (i) includes all three women;

   (ii) includes all three women and has one of the women as chairperson.

5. Suppose a typical situation during dinner hour at the college cafeteria is that 50 students are vying for ten empty seats. On a given night during dinner hour, if five of the 50 students are kinesiology majors, find an expression for (do not evaluate) the probability that exactly four of the kinesiology majors get seats.

6. In the Virginia lottery's game of "Lotto," six numbers are selected at random, without replacement, from the numbers 1 through 44. Find an expression for (do not evaluate) the probability of matching:

   (i) All six numbers;

   (ii) Exactly five of the six numbers;

   (iii) Exactly four of the six numbers.

7. At NCAA football games, each team is required to give the game officials ten game balls for use in each game. The teams use the same brand and type of ball, but all teams brand the balls with the school's name. After the game the balls which are used are given away by the NCAA for charity auctions and other uses. The left-overs are returned to the school whose name was attached. When University of Virginia played William and Mary in a game, only five of the 20 balls were used. At the end of the game nine balls were returned to University of Virginia and six to William and Mary. If the balls were selected at random, what is the probability of this outcome?

8. The chairperson of the mathematics department is forced, by declining enrollments, to lay off three of her teachers. Eleven of the twenty teachers in the department are female. If the three to be laid off are chosen at random from the twenty teachers, find an expression for (do not evaluate) the probability that more males than females are laid off.

9. A hospital ward contains 15 male and 20 female patients. Five patients are chosen at random to receive a special treatment. Find an expression for (do not evaluate) the probability that at least two patients of each sex are selected.

10. Three chips are drawn at random without replacement from a box containing ten chips numbered 1, 2, 3, ..., 10. Find an expression for (do not evaluate) the probability that the largest number drawn is six.

11. Ten tires of different brands are ranked from one to ten (best to worst) according to mileage performance. If four of these tires are chosen at random by a customer, find an expression for the probability that the best tire among those selected by the customer is actually ranked third among the original ten. Do not evaluate the expression.

12. Four chips are drawn without replacement from a box containing ten chips numbered $1, 2, 3, ..., 10$.

   (i) Find an expression for (do not evaluate) the probability that the second smallest number drawn is four.

(ii) The range of the set of values is the largest number minus the smallest number. Find an expression for (do not evaluate) the probability that the range of the numbers drawn equals six.

13. In Example 5.5, verify that $E(X) = 1$ for all three sampling plans.

14. In an election, the challenger, who won, had a margin of victory of four votes. Among the 120 votes cast were seven invalid votes. Should the incumbent challenge the election results?

15. Urn I contains four red chips and three white chips. Urn II contains four red chips and five white chips. Find an expression for (do not evaluate) the probability of getting two chips of each color if:

(i) Two chips are drawn at random, without replacement, from each urn.

(ii) An urn is selected at random and then four chips drawn at random, without replacement, from that urn.

16. The United States Senate has 100 members. A random sample of two senators is taken, and it is found that both of them are engaged in corruption or fraud of some kind.

(i) If there are c corrupt senators, where $c > 1$, what is the probability that a sample of two senators selected at random will contain two corrupt senators? Answer in terms of c.

(ii) Use your calculator to find the value of c, by trial and error, which satisfies the following statement: "The observation of two corrupt senators, in a sample of size two selected at random, is a rare occurrence (probability less than .05) unless there are at least c corrupt senators."

17. We want to estimate the number of fish in a pond. We catch three fish at random (without replacement), tag them, and then release them back into the pond. A day later we catch three fish and we observe that two of the three are tagged. Estimate the number of fish in the pond.

18. In a certain country the draft-status priorities of eligible men are determined by their birthdays. Suppose the numbers 1 to 31 are assigned to men with birthdays from January 1 to January 31, numbers 32 to 60 to men with birthdays from February 1 to February 29, etc. Then numbers are selected at random, without replacement,

from 1 to 366 until all of them are chosen. Those with birthdays corresponding to the first number drawn would have the highest draft priority, those with birthdays corresponding to the second number drawn would have the second-highest draft priority, and so on. Let X denote the largest of the first five numbers selected.

(i) What possible values can x assume?

(ii) Find an expression for (do not evaluate) $P(X = 10)$.

(iii) Suppose $A =$ "those with the five highest draft priorities were all born in January." Write the event A in terms of an expression involving X.

19. Urn I contains three white and three red chips. Urn II contains five white and one red chip. An urn is selected at random and then three chips are selected at random, without replacement, from the urn. If two of the three selected chips are white, what is the probability that the urn selected was urn II?

20. A man plays the Virginia lottery's game of "Lotto" (refer to Problem 6 in this chapter) once a week for 50 years. Find the probability that he matches all six numbers at least once.

21. In a community of 18 potential voters, ten are for legalizing marijuana and eight are against it. Suppose that a vote is taken to determine the will of the majority. If five random persons of these 18 potential voters do not vote, find an expression for (do not evaluate) the probability that those against legalizing marijuana will win.

22. In the fall of 2001, anthrax was discovered in a mailroom which receives mail from a U.S. Postal Distribution Center in Brentwood, MD. This Center also distributes mail to approximately 3200 other mailrooms in Washington, DC and the surrounding area. After the anthrax was found in this mailroom, a sample of mailrooms were selected at random for anthrax testing from the other 3200 mailrooms since there was concern that some of these other mailrooms had also been contaminated. It was not feasible to test all 3200 mailrooms. The objective was to estimate how many of the other 3200 mailrooms had been contaminated.

(i) Why does this scenario belong in Chapter 5? Explain your answer.

(ii) Which example in this chapter best provides a working model for this scenario? Explain your answer.

23. In Example 5.8, assume the drug wears off over a two-day period but the user tests positive with certainty (probability of one) during the wear-off period. Find P(user tests $+$).

24. In Example 5.8, assume the drug wears off over a two-day period but the user tests positive with certainty during the wear-off period.

   (i) Show that

   $$P(\text{user tests negative}) = P(\text{user tests } -) = \frac{28}{30} \cdot \frac{27}{29} \cdot \frac{26}{28}.$$

   (ii) Find P(user tests $+$) from (i).

   (iii) Compare your answer in (ii) with the answer to the previous problem.

6

Sampling with Replacement

This chapter deals with distributions which model many real-life situations. We assume a series of trials where there are two possible outcomes per trial and the probability for each outcome is the same for all trials. We also assume that outcomes of different trials are independent events. Applications abound in law, medicine, political science, sports, and other areas.

6.1   Binomial Model

The following example is identical to Example 5.2 except that sampling is carried out *with* replacement.

Example 6.1. A box contains eight red and six white chips. Four chips are drawn with replacement from the box. Let X denote the number of red chips drawn. Find an expression for the probabilities of the following events:

(i) Three red chips followed by one white chip.

(ii) One red chip followed by one white chip and then two red chips.

(iii) $X = 3$

(iv) $X \geq 3$

Solution:

(i) $P(RRRW) = \frac{8}{14} \cdot \frac{8}{14} \cdot \frac{8}{14} \cdot \frac{6}{14} = (\frac{8}{14})^3 (\frac{6}{14})^1$

(ii) $P(RWRR) = \frac{8}{14} \cdot \frac{6}{14} \cdot \frac{8}{14} \cdot \frac{8}{14} = (\frac{8}{14})^3 (\frac{6}{14})^1$

(iii) $\begin{aligned}
P(X = 3) \ &= \ P(RRRW) + P(RRWR) \\
&\quad + P(RWRR) + P(WRRR) \\[2mm]
&= \ \left(\frac{8}{14}\right)^3 \left(\frac{6}{14}\right)^1 + \left(\frac{8}{14}\right)^3 \left(\frac{6}{14}\right)^1 \\
&\quad + \left(\frac{8}{14}\right)^3 \left(\frac{6}{14}\right)^1 + \left(\frac{8}{14}\right)^3 \left(\frac{6}{14}\right)^1 \\[2mm]
&= \ 4 \left(\frac{8}{14}\right)^3 \left(\frac{6}{14}\right)^1 \\[2mm]
&= \ \frac{4!}{3!1!} \left(\frac{8}{14}\right)^3 \left(\frac{6}{14}\right)^1
\end{aligned}$

since the numbers of terms in the sum equals the number of words with $3R$'s, $1W$ which is $\frac{4!}{3!1!} = 4$.

(iv) $\begin{aligned}
P(X \geq 3) \ &= \ P(X = 3) + P(X = 4) \\[2mm]
&= \ \frac{4!}{3!1!} \left(\frac{8}{14}\right)^3 \left(\frac{6}{14}\right)^1 + \frac{4!}{4!0!} \left(\frac{8}{14}\right)^3 \left(\frac{6}{14}\right)^0 .
\end{aligned}$

The difference between this model and the one used in Example 5.2 is that we have independence from draw to draw here since we are sampling with replacement. So events such as "red chip on first draw" and "red chip on second draw" are independent events.

In terms of an experiment where chips are drawn from a box, the binomial model satisfies three conditions:

(i) Two types of chips in the box.

(ii) Fixed numbers of draws from the same box.

(iii) Independence from draw to draw.

Example 6.2. Suppose 60% is a passing grade on a test. If there are five questions on the test and a student guesses on every question on the test, what is the probability that she passes?

(i) Assume a true-false test.

(ii) Assume a multiple choice test with four alternatives for each question.

Solution:

(i) Since there are five questions on the test, a passing grade of 60% corresponds to getting at least three questions correct. Let C denote a correct answer and I an incorrect answer. For a true-false test, $P(C) = .5$ and $P(I) = .5$.

$$P(\text{she passes}) = P(3C, 2I) + P(4C, 1I) + P(5C)$$

$$= \frac{5!}{3!2!}(.5)^3(.5)^2 + \frac{5!}{4!1!}(.5)^4(.5)^1 + (.5)^5 = .5.$$

For the binomial model, we think of drawing five chips with replacement from a box containing one C and one I. Each draw corresponds to one question and the probability of getting a C(correct answer) is always $\frac{1}{2} = .5$.

(ii) Since she is guessing and there are four alternatives for each question, $P(C) = \frac{1}{4} = .25$ and $P(I) = .75$.

$$P(\text{she passes}) = P(3C, 2I) + P(4C, 1I) + P(5C)$$

$$= \frac{5!}{3!2!}(.25)^3(.75)^2 + \frac{5!}{4!1!}(.25)^4(.75)^1 + (.25)^5$$

$$= .0668.$$

For the binomial model, we think of drawing 5 chips with replacement from a box containing one C and three I's. Each draw corresponds to one question and the probability of getting a C(correct answer) is always $\frac{1}{4} = .25$.

Example 6.3. Consider a series of games played by two teams, A and B. Suppose that the team that wins the majority of the games in the series, wins the series. In addition, suppose that team A is twice as good as team B; that is $P(A$ wins a game$) = 2/3$ and $P(B$ wins a game$) = 1/3$. Also, assume independence of outcomes from game to game. One reason for playing a series of games rather than a single game (in addition to financial reasons) is to increase the chances that the better team will win.

Consider a three-game series:

Our model for this example is a box filled with three chips, two of which are labelled A and one which is labelled B. Three chips are drawn with replacement from this box. By drawing with replacement from this box, the probability of getting an A is always $\frac{2}{3}$. Each draw corresponds to a game in the series.

$$P(A \text{ wins the series}) = P(A \text{ wins at least two of the three games})$$

$$= P(A \text{ wins 2 games}) + P(A \text{ wins 3 games})$$

$$= \tfrac{3!}{2!1!} \left(\tfrac{2}{3}\right)^2 \left(\tfrac{1}{3}\right)^1 + \left(\tfrac{2}{3}\right)^3 = .74.$$

Consider a five-game series:

$$P(A \text{ wins the series})$$

$$= P(A \text{ wins at least three of the five games})$$

$$= P(A \text{ wins 3 games}) + P(A \text{ wins 4 games}) + P(A \text{ wins 5 games})$$

$$= \frac{5!}{3!2!} \left(\frac{2}{3}\right)^3 \left(\frac{1}{3}\right)^2 + \frac{5!}{4!1!} \left(\frac{2}{3}\right)^4 \left(\frac{1}{3}\right)^1 + \left(\frac{2}{3}\right)^5 = .79.$$

Note that $\frac{5!}{5!0!} \left(\frac{2}{3}\right)^5 \left(\frac{1}{3}\right)^0 = \left(\frac{2}{3}\right)^5$.

So the chances of team A winning a single game are .67. With a three game series, A's chance of winning the series increases to .74. With a five game series these chances are further increased to .79. Even though A is twice as good as B, the chances of B winning a five game series are still better than one in five (the probability is .21).

Suppose we wanted to be at least 90% certain that A would win the series. What is the minimum number of games to be played to satisfy that condition? By a trial-and-error solution, it is possible to show that a fifteen game series would be needed.

Example 6.4. (Overbooking.) A certain commuter airline company has planes that will carry six passengers. From past experience they have found that one-third of the people who make reservations are "no shows." Find the largest number of reservations they can take for a flight if they want to be at least 80% sure that everyone who makes a reservation gets a seat on the plane.

Solution: First we interpret the statement that one-third of the people who make reservations are "no shows." Does this mean that if six people make reservations that two will be "no shows"? No. The quantity $\frac{1}{3}$ is a probability or a long-run proportion. The number of "no shows" for a certain flight could be any integer from 0 to 6 when six people make reservations. Actually, the expected number of "no shows" turns out to be two. We could think in terms of a simulation using the random number target where the numbers 1 and 2 correspond to a person being a "no show" with probability $\frac{2}{6} = \frac{1}{3}$. Now, on to the question at hand. How many reservations should the company take if they want to be at least 80% sure that everyone who makes a reservation gets a seat on the plane? The solution will be by trial and error.

Suppose they take six reservations:
Since the plane will carry six passengers,

$$P \text{ (everyone gets a seat)} = 1.$$

Suppose they take seven reservations:
Now, there is a chance that all seven people show up and therefore not everyone will get a seat. For our model we imagine a box with three chips; two of the chips are marked S for "show" and one of the chips is marked N for "no show." We then think of drawing seven chips with replacement from the box. Each draw corresponds to one of the seven people who will show (chip S) with probability $\frac{2}{3}$ or be a "no show" (chip N) with probability $\frac{1}{3}$. Do not confuse the contents of the box with the number of draws. That is, we have $2S$ and $1N$ chips in the box because the probability of an N(no show) must be $\frac{1}{3}$ for each person. This will be the case regardless of the number of reservations taken.

$$P(\text{everyone gets a seat}) = P(\text{at most 6 people show}).$$

Rather than calculating $P(\text{at most 6 people show}) = P(0 \text{ people show}) + P(1 \text{ person shows}) + \cdots + P(6 \text{ people show})$ it is easier to consider the complementary event that seven people show.

$$
\begin{aligned}
P(\text{everyone gets a seat}) \quad &= \quad P(\text{at most 6 people show}) \\[6pt]
&= \quad 1 - P(7 \text{ people show}) \\[6pt]
&= \quad 1 - P(7S \text{ chips}, 0N \text{ chips}) \\[6pt]
&= \quad 1 - \frac{7!}{7!0!}\left(\frac{2}{3}\right)^{7}\left(\frac{1}{3}\right)^{0} \\[6pt]
&= \quad .94.
\end{aligned}
$$

Suppose they take eight reservations:

Again our box contains 2 S chips and 1 N chip, but now we draw eight chips with replacement to correspond to the eight people making reservations.

$$
\begin{aligned}
P(\text{everyone gets a seat}) &= P(\text{at most 6 people show}) \\
&= 1 - P(\text{7 or 8 people show}) \\
&= 1 - [P(7S, 1N) + P(8S, 0N)] \\
&= 1 - \left[\frac{8!}{7!1!} \left(\frac{2}{3}\right)^7 \left(\frac{1}{3}\right)^1 + \frac{8!}{8!0!} \left(\frac{2}{3}\right)^8 \left(\frac{1}{3}\right)^0 \right] \\
&= .805.
\end{aligned}
$$

Suppose they take nine reservations:

Using the same reasoning, one can show that

$$P(\text{everyone gets a seat}) = .622.$$

Reviewing our results to this point:

- If they take 6 reservations, P(everyone gets a seat) = 1.

- If they take 7 reservations, P(everyone gets a seat) = .94.

- If they take 8 reservations, P(everyone gets a seat) = .805.

- If they take 9 reservations, P(everyone gets a seat) = .622.

Thus the largest number of reservations they can take for a certain flight if they want to be at least 80% sure that everyone gets a seat is eight.

Our model assumes that whether or not one person is a "no show" does not affect the probability of another person being a "no show." When we sample with replacement, independence is built into our model.

We continue Example 6.4 by using a different criteria for determining the number of reservations the airline company should take.

Example 6.5. (Overbooking continued.)   A certain commuter airline company has planes that will carry six passengers. From past experience they have found that one-third of the people who make reservations are "no shows." Suppose the price of a ticket is $1. If people are bumped because there are not enough seats to accommodate the people who have shown up they can travel on a future flight free of charge. So the company loses $1 per flight for each bumped person because they must give that person a free ticket or equivalently $1 for a future flight. How many reservations should the company take if they want the expected gain to the company for that flight to be maximum? Again we assume that whether or not one person is a "no show" does not affect the probability of another person being a "no show."

Solution: It is clear that the expected gain to the company for that flight will be greater if the company takes six reservations than if they take fewer than six, so we will start off by assuming that they take six reservations. The solution will be by trial and error.

Suppose they take six reservations:

$$
\begin{aligned}
E(\text{gain}) &= \quad 1 \cdot P(1 \text{ person shows}) + 2 \cdot P(2 \text{ persons show}) \\
&\quad + \cdots + \; 6 \cdot P(6 \text{ persons show}) \\
&= \quad 1 \cdot P(1S, 5N) + 2 \cdot P(2S, 4N) \\
&\quad + \cdots + 6 \cdot P(6S, 0N) = \$4.00,
\end{aligned}
$$

after substituting $\frac{6!}{1!5!} \left(\frac{2}{3}\right)^1 \left(\frac{1}{3}\right)^5$ for $P(1S, 5N)$, etc., and then summing. We could have included the term $0 \cdot P(0 \text{ persons show})$ but that product is 0.

Suppose they take seven reservations:

Now if seven people show, then one person gets bumped and the company will get $1 for each of the six people who are seated but will lose $1 for the bumped person. So the gain to the company for that flight is $6 − $1 = $5.

$$
\begin{aligned}
E(\text{gain}) &= \quad 1 \cdot P(1 \text{ person shows}) \\
&\quad + \cdots + \; 6 \cdot P(6 \text{ persons show}) + 5 \cdot P(7 \text{ persons show}) \\
&= \quad 1 \cdot P(1S, 6N) + 2 \cdot P(2S, 5N) \\
&\quad + \cdots + \; 6 \cdot P(6S, 1N) + 5 \cdot P(7S, 0N) = \$4.55.
\end{aligned}
$$

Suppose they take eight reservations:

Now if seven people show, then one person gets bumped and the company will get $1 for each of the six people who are seated but will lose $1 for the bumped person. So the gain is $6 - $1 = $5. If eight people show, then two people get bumped and the company will get $1 for each of the six people who are seated but will lose $1 for each of the two bumped people. So the gain to the company for that flight is $6 - $2 = $4.

$$E(\text{gain}) \quad = \quad 1 \cdot P(1 \text{ person shows})$$

$$+ \cdots + \; 5 \cdot P(7 \text{ persons show}) + 4 \cdot P(8 \text{ persons show})$$

$$= \quad 1 \cdot P(1S, 7N) + 2 \cdot P(2S, 6N) + \cdots + 6 \cdot P(6S, 2N)$$

$$+ \; 5 \cdot P(7S, 1N) + 4 \cdot P(8S, 0N) = \$4.87.$$

Using the same reasoning, we can show that if they take nine reservations $E(\text{gain}) = \$4.91$.

Reviewing the results to this point and including additional results:

- If they take 5 reservations, $E(\text{gain}) = \$3.33$.

- If they take 6 reservations, $E(\text{gain}) = \$4.00$.

- If they take 7 reservations, $E(\text{gain}) = \$4.55$.

- If they take 8 reservations, $E(\text{gain}) = \$4.87$.

- If they take 9 reservations, $E(\text{gain}) = \$4.91$.

- If they take 10 reservations, $E(\text{gain}) = \$4.71$.

- If they take 11 reservations, $E(\text{gain}) = \$4.32$.

As the number of reservations increases, $E(\text{gain})$ increases and then decreases. To maximize the expected gain, they should take nine reservations.

Note that if the cost of a ticket had been $1000 rather than $1, each of the probability multipliers in $E(\text{gain})$ would be multiplied by 1000 and so $E(\text{gain})$ would be multiplied by 1000. So the ticket cost change would affect $E(\text{gain})$ but would not affect the number of reservations needed to maximize the expected gain because all the $E(\text{gain})$ values would be multiplied by 1000.

We see that the answer to Example 6.5 differs from the answer to Example 6.4. This is because we are using different criteria.

Example 6.6. (Rescue Attempt of U.S. Hostages in Iran in 1980.) In the spring of 1980, after diplomatic efforts to free 53 American hostages in Iran had failed, President Carter approved a daring mission in which a special unit from the Air Force would attempt to rescue these hostages. The helicopters involved in this mission were designed to carry in the special unit and bring them out with the hostages. It was a long trip and the helicopters would be extended to their limits. Therefore, the Pentagon decided to add two extra helicopters to the minimum of six that would be required to carry out the special unit and the hostages. What happened was that three of the eight helicopters malfunctioned and the mission had to be canceled in progress. Several American servicemen died in the effort and the 53 American hostages were subsequently dispersed rather than being kept in one location. Afterwards, some analysts argued that more helicopters should have been assigned to the mission while others disagreed saying that more helicopters in the air would have increased the chances of them being spotted and shot down.

We base our probability model on the previous information. The Pentagon estimated that the hostages would be rescued if at least six helicopters were successful. As a safety factor they decided to send two extra helicopters. How would the chance of a successful mission change if they had sent three extra helicopters rather than two? The Pentagon estimated that if eight helicopters were sent, the probability that at least six were successful (and the hostages would be rescued) was .965. What is the probability that the hostages would have been rescued if they had sent in nine helicopters rather than eight?

Solution: The estimate of the probability that at least six of eight helicopters are successful is .965. We must find the probability that the hostages would be rescued if nine helicopters were sent which is the probability that at least six of nine helicopters are successful.

Let $p = P(\text{any helicopter is successful}) = P(S)$.

Then $P(\text{any helicopter is not successful}) = P(N) = 1 - p$.

$$.965 = P(\text{at least 6 of 8 helicopters are successful})$$

$$.965 = P(6S, 2N) + P(7S, 1N) + P(8S)$$

$$.965 = \frac{8!}{6!2!}p^6(1-p)^2 + \frac{8!}{7!1!}p^7(1-p) + p^8.$$

The trial-and-error solution to the last equation is $p = .903$. This is arrived at by plugging different values of p into this equation until the equation is satisfied. This gives us the probability that any helicopter is successful.

We want the probability that at least six of the nine helicopters are successful.

P(at least 6 of 9 helicopters are successful)

$$= \quad P(6S, 3N) + P(7S, 2N) + P(8S, 1N) + P(9S)$$

$$= \quad \frac{9!}{6!3!}(.903)^6(.097)^3 + \frac{9!}{7!2!}(.903)^7(.097)^2$$

$$+ \frac{9!}{8!1!}(.903)^8(.097)^1 + (.903)^9$$

$$= \quad .992.$$

So sending an additional three helicopters rather than an additional two would have increased the probability that the hostages would be rescued from 96.5% to 99.2%.

The following example, which combines the binomial model and conditional probabilities, lays the foundation for Example 6.8.

Example 6.7. Roll a balanced die five times. Let A = "more than one 6" and B = "at least one 6." Find $P(A|B)$.

Solution: Since we are concerned with whether or not each outcome is a 6, we take as our model a box filled with five chips marked N (not a 6) and one chip marked S (six). Now if we draw with replacement from this box, we get an S chip (six) with probability $\frac{1}{6}$. We must draw five chips, each draw corresponding to the roll of a balanced die.

$$A \quad = \quad \text{"2, 3, 4, or 5 } S \text{ chips"}$$
$$B \quad = \quad \text{"1, 2, 3, 4, or 5 } S \text{ chips"}$$

So, "A and B" = "2,3,4, or 5 S chips." Remember that A and B includes those outcomes in A and also in B; that is, outcomes common to both events.

In this example, "A and B" = A but in general this will not be true.

$$P(A|B) \quad = \quad \frac{P(A \text{ and } B)}{P(B)}$$

$$= \quad \frac{P(2, 3, 4 \text{ or } 5S \text{ chips})}{P(1, 2, 3, 4, \text{ or } 5S \text{ chips})}.$$

For both numerator and denominator, it will be easier to consider the complementary events.

$$P(A|B) = \frac{P(2, 3, 4, \text{ or } 5S \text{ chips})}{P(1, 2, 3, 4, \text{ or } 5S \text{ chips})}$$

$$= \frac{1 - P(0 \text{ or } 1S \text{ chips})}{1 - P(0S \text{ chips})}$$

$$= \frac{1 - [P(0S, 5N \text{ chips}) + P(1S, 4N \text{ chips})]}{1 - P(0S, 5N \text{ chips})}$$

$$= \frac{1 - \left[\frac{5!}{0!5!}\left(\frac{1}{6}\right)^0\left(\frac{5}{6}\right)^5 + \frac{5!}{1!4!}\left(\frac{1}{6}\right)^1\left(\frac{5}{6}\right)^4\right]}{1 - \frac{5!}{0!5!}\left(\frac{1}{6}\right)^0\left(\frac{5}{6}\right)^5}$$

$$= .328.$$

Example 6.8. ("People vs. Collins.") In 1964, the purse of an elderly woman shopping in Los Angeles was snatched by a young white female wearing a blond ponytail. The thief fled on foot but was seen shortly thereafter getting into a yellow car driven by a black male who had a mustache and a beard. A police investigation subsequently turned up a suspect, one Janet Collins, who was blond, wore a ponytail, and lived with a black man who drove a yellow car and had a mustache and a beard. An arrest was made. Having no witnesses who could identify the party, the prosecution's case rested on the unlikelihood that a given couple would match all six characteristics reported by several eyewitnesses to the crime. None of these eyewitnesses were able to identify the suspects. It was estimated that the joint occurrence of a white female with blond hair tied in a ponytail riding in a yellow car with a black man having a beard and a mustache was on the order of one in twelve million—a number so small the prosecution argued that Ms. Collins and her male friend were guilty. The jury agreed and handed down a verdict of second-degree robbery.

Characteristic	Estimated Probability
Yellow car	.1
Man with mustache	.25
Woman with ponytail	.1
Woman with blond hair	.33
Black man with beard	.1
Interracial couple in car	.001

P(couple has all 6 characteristics)

$$= \quad (.1)(.25)(.1)(.33)(.1)(.001) = (\tfrac{1}{12,000,000})$$

assuming independence.

The Supreme Court of California later disagreed and reversed the decision based on an argument using the binomial model.

Let H = "couple has all 6 characteristics," and
N = "couple does not have all 6 characteristics."

Also, let M represent the number one million. Assume there were $2,000,000 = 2M$ couples in the Los Angeles area who could have committed the crime. Also, assume $P(H) = \tfrac{1}{12M}$ which was the figure arrived at by the prosecutor.

For the model, we have a box filled with $12M$ chips of which one chip is marked H and the other chips are marked N. We think of drawing $2M$ chips from this box with replacement, each chip corresponding to a couple. So for each couple, $P(H) = \tfrac{1}{12M}$. We know there is at least one couple in the Los Angeles area having all six characteristics. The question is: What is the probability that there is at least one other couple in the Los Angeles area with all six characteristics?

If we let $A =$"more than one couple have all six characteristics" and $B =$"at least one couple have all six characteristics," then the problem is to find $P(A|B)$. We are given B because of the existence of Janet Collins and her male friend.

The reasoning involved in finding $P(A|B)$ is much like that used in Example 6.7. Here, "A and B" $= A$. So

$$P(A|B) = \frac{P(A \text{ and } B)}{P(B)}$$

$$= \frac{P(2,3,4,\ldots, \text{ or } 2M \text{ chips marked } H)}{P(1,2,3,4,\ldots, \text{ or } 2M \text{ chips marked } H)}$$

$$= \frac{1 - P(0 \text{ or } 1 \text{ chips marked } H)}{1 - P(0 \text{ chips marked } H)}$$

$$= \frac{1 - [P(0H \text{ chips}, (2M)N \text{ chips }) + P(1H \text{ chip}, (2M-1)N \text{ chips})]}{1 - P(0H \text{ chips}, (2M)N \text{ chips})}$$

$$= \frac{1 - \left[\frac{(2M)!}{(2M)!0!}\left(\frac{1}{12M}\right)^0\left(\frac{12M-1}{12M}\right)^{2M} + \frac{(2M)!}{(2M-1)!1!}\left(\frac{1}{12M}\right)^1\left(\frac{12M-1}{12M}\right)^{2M-1}\right]}{1 - \frac{(2M)!}{(2M)!0!}\left(\frac{1}{12M}\right)^0\left(\frac{12M-1}{12M}\right)^{2M}}$$

$$= \quad .08.$$

Thus there is a reasonable chance (.08) that there was at least one other couple in the area having all six characteristics. On that basis, the Supreme Court of California overturned the original decision of the lower court.

Example 6.9. (Modeling Jury Verdicts.) This example deals with the effect of jury size on the probability of conviction in a jury trial. The model under consideration here is known as the modified Walbert model. We let $p = P(\text{juror initially votes guilty})$.

In the modified Walbert model, if the initial jury vote is not unanimous, the jurors will deliberate further and the majority will *usually* prevail. The larger the majority, the more often it will prevail. In a jury with n members, a group of size k voting for conviction in the initial vote would convince the rest of the jury k/n of the time. For example in a three-person jury, two jurors initially voting guilty (G) would lead to a guilty jury verdict (conviction) 2/3 of the time and a not-guilty jury verdict 1/3 of the time. For a three-person jury, one juror initially voting guilty would lead to a guilty jury verdict 1/3 of the time and a not-guilty jury verdict 2/3 of the time. For a five-person jury, three jurors initially voting guilty would lead to a guilty jury verdict 3/5 of the time.

Suppose we have a three-person jury:

Using the addition and multiplication rules,

$$P(\text{conviction})$$

$$= \quad P(1 \text{ jurors initially votes } G \text{ and conviction})$$

$$+ \, P(2 \text{ jurors initially vote } G \text{ and conviction})$$

$$+ \, P(3 \text{ jurors initially vote } G \text{ and conviction})$$

$$= \quad P(1 \text{ juror initially votes guilty})$$

$$\cdot P(\text{conviction}|1 \text{ juror initially votes guilty})$$

$$+ P(2 \text{ jurors initially vote guilty})$$

$$\cdot P(\text{conviction}|2 \text{ jurors initially vote guilty})$$

$$+ P(3 \text{ jurors initially vote guilty})$$

$$\cdot P(\text{conviction}|3 \text{ jurors initially vote guilty}).$$

Substituting into the last equation,

$$P(\text{conviction}) \quad = \quad \left[\frac{3!}{1!2!} p^1 (1-p)^2 \right] \frac{1}{3} + \left[\frac{3!}{2!1!} p^2 (1-p)^1 \right] \frac{2}{3} + \left[p^3 \right] 1$$

$$= \quad \left[3p(1-p)^2 \right] \frac{1}{3} + 3 \left[p^2 (1-p) \right] \frac{2}{3} + p^3$$

$$= \quad p \left(1 - 2p + p^2 \right) + 2p^2 (1-p) + p^3$$

$$= \quad p - 2p^2 + p^3 + 2p^2 - 2p^3 + p^3$$

$$= \quad p.$$

It is possible to show that $P(\text{conviction}) = p$ for all jury sizes. Therefore, for this model the size of the jury does not affect the probability of conviction. A problem at the end of this chapter asks you to verify that $P(\text{conviction}) = p$ for a two-person jury.

Consider a class exercise. The following three pages give three actual cases where jury members must decide on the guilt or innocence of a defendant. Read through each case and decide whether the defendant is guilty (G) or not guilty (N). Then for each case, divide the class into small student juries and discuss the case reaching a guilty or not-guilty decision for your jury. Summarize the data from the jury vote. What percentage of the juries voted in a way that is consistent with the initial vote of the jurors? For example, suppose you selected five-person juries. Of those juries where the initial jury vote was 3-G, 2-N, what percentage of the juries voted guilty? Is it approximately 60%? If the percentages are not at all consistent with the modified Walbert model, propose some variation of this model which is suggested by the data. Then investigate how $P(\text{conviction})$ depends on jury size for the model that you propose.

Alex and Brett were unhappy youths. They made a suicide pact. To carry out the pact, Alex drove his car over a cliff, with Brett as his passenger. Brett died. Alex recovered. Alex is charged with murder.

Is Alex guilty?

Reprinted with permission of Sterling Publishing Co., NY, NY from *WHAT'S THE VERDICT?* by Ted I Valliant & Marcel Theroux, Illustrated by Myron Miller, ©1991 by Ted Le Valliant & Marcel Theroux

Figure 6.1. Case 1.

Pierre had to drive long distances because of his work. He liked to use the automatic cruise control. He set the control at the speed limit but was nevertheless ticketed for speeding. At his trial, he showed that the automatic cruise control had malfunctioned on the day in question.

Is Pierre guilty of speeding?

Figure 6.2. Case 2.

Mildred stabbed the victim. The victim was transported to the hospital. The victim was twice dropped by the person carrying him. The hospital had no facilities for blood transfusions. Had the victim received blood, he would have had a 75 percent chance of survival. Mildred is charged with murder.
  Is Mildred guilty?

Reprinted with permission of Sterling Publishing Co., NY, NY from *WHAT'S THE VERDICT?* by Ted Valliant & Marcel Theroux, Illustrated by Myron Miller, ©1991 by Ted Le Valliant & Marcel Theroux

Figure 6.3. Case 3.

Example 6.10. (Blood Testing.) Suppose 30 people are given a blood test for a disease. Assume that each person has probability .0105 of having the disease independent of the others. To examine each person's blood individually means a total of 30 tests will be run which may be inefficient. Another strategy is to pool the 30 blood samples into three groups of size ten and run the blood test on each group. If a group test is negative, all ten people in that group are free from the disease. So only one test is necessary for that group. If a group test is positive (meaning at least one person in the group has the disease), each person's blood is then tested separately resulting in a total of 11 tests being done for those ten persons. Find the expected number of tests when pooling and compare it to 30. Then find the optimal group size (the group size that minimizes the expected number of tests).

Solution: Let X denote the total number of tests for all groups. If all three groups are negative then $X = 3$. If two groups are negative and one is positive then $X = 1 + 1 + 11 = 13$ since there would be one test for each negative group and 11 tests for the positive group. If one group is negative and two groups are positive, then $X = 1 + 11 + 11 = 23$. If all three groups are positive, then $X = 11 + 11 + 11 = 33$. So the possible values assumed by X are 3, 13, 23, 33. In order to determine $E(X)$ we must first find $P(X = x)$ for each possible value assumed by X.

$$
\begin{aligned}
&P(\text{any group is negative}) \\
&= \quad P(10 \text{ persons are negative}) \\
&= \quad P(\text{person 1 is negative and person 2 is negative} \\
&\qquad \text{and} \ldots \text{person 10 is negative}) \\
&= \quad P(\text{person 1 is negative})P(\text{person 2 is negative}) \\
&\qquad \ldots P(\text{person 10 is negative}) \\
&= \quad (1 - .0105)^{10} \\
&= \quad .9 \text{ assuming independence.}
\end{aligned}
$$

Thus,

P(any group is positive) $= 1 - P$(any group is negative)$= 1 - .9 = .1$.

Let $P = $ "positive group" and $N = $ "negative group." Then our model consists of a box with nine chips marked N and one chip marked P. This ensures that P(any group is positive) $= .1$. We draw three chips from this box where each chip corresponds to a group.

$$
\begin{aligned}
P(X = 33) &= \quad P(3 \text{ positive groups}) \\
&= \quad P(3P \text{ chips}, 0N \text{ chips}) \\
&= \quad \frac{3!}{3!0!}(.1)^3(.9)^0 \\
&= \quad .001.
\end{aligned}
$$

$$P(X = 23) \ = \ P(2 \text{ positive groups})$$
$$= \ P(2P \text{ chips}, 1N \text{ chip})$$
$$= \ \frac{3!}{2!1!}(.1)^2(.9)^1$$
$$= \ .027.$$

$$P(X = 13) \ = \ P(1 \text{ positive groups})$$
$$= \ P(1P \text{ chip}, 2N \text{ chips})$$
$$= \ \frac{3!}{1!2!}(.1)^1(.9)^2$$
$$= \ .243.$$

$$P(X = 3) \ = \ P(0 \text{ positive groups})$$
$$= \ P(0P \text{ chips}, 3N \text{ chips})$$
$$= \ \frac{3!}{0!3!}(.1)^0(.9)^3$$
$$= \ .729.$$

So the distribution of X is:

x	$P(X = x)$
3	.729
13	.243
23	.027
33	.001

$$E(X) = 3(.729) + 13(.243) + 23(.027) + 33(.001) = 6.000.$$

The expected number of tests when pooling with groups of size ten is 6.000 compared to 30 tests without pooling.

Now determine $E(X)$ for all possible group sizes to find the optimal group size. That is, which group size will minimize the average amount of testing required. Would it be better to have three groups with ten persons per group or 15 groups with two persons per group if our goal is to minimize the expected number of tests? The table on the next page gives us the answer.

Number of Groups	Number of Persons per Group	$E(X)$
1	30	9.14
2	15	6.39
3	10	6.00
5	6	6.84
6	5	7.54
10	3	10.93
15	2	15.62

We have already worked out $E(X)$ for a group size of ten and you will be asked to work out $E(X)$ for a group size of 15 as a Chapter problem. From the table we see that the optimal group size is ten since this gives the smallest value of $E(X)$ which is 6.000.

⋆ 6.2 Approximating Sums of Binomial Probabilities

Binomial sums are tedious to evaluate computationally when the sample size is large. Fortunately there are good approximations to binomial sums. In this section, we present an approximation which applies when the sample size is large (larger than 40) and the two possible outcomes are equally likely. We state the approximation and then give an example. The reason why the approximation works is beyond the level of this text.

Let n denote the sample size and X denote our random variable for the binomial model.

The following gives us an approximation for $P(X \geq x)$ when $x > \frac{n}{2}$:

If $x > \frac{n}{2} + 1.1\sqrt{n}$, then $P(X \geq x) \leq .01$.

If $\frac{n}{2} < x \leq \frac{n}{2} + 1.1\sqrt{n}$,

$$\text{then } P(X \geq x) \sim \frac{1}{2} - \frac{1}{10}\left[\left(\frac{2x-n}{\sqrt{n}}\right)\left\{4.4 - \left(\frac{2x-n}{\sqrt{n}}\right)\right\}\right],$$

where the symbol $\sim$ denotes "approximately equal to."

So whether we use the bound of .01 or the formula to approximate $P(X \geq x)$ depends on which of the inequalities on x is satisfied. If the value of x is in the interval between $\frac{n}{2}$ and $\frac{n}{2}+1.1\sqrt{n}$ then we approximate $P(X \geq x)$ from the formula. On the other hand if x is greater than $\frac{n}{2} + 1.1\sqrt{n}$ then we give .01 as the upper bound on $P(X \geq x)$.

Example 6.11. Marni takes a 100-question true-false exam. If she guesses on every question, find the approximate probability that she gets:

(i) at least 55 questions correct;

(ii) at least 60 questions correct;

(iii) at least 62 questions correct.

Solution: Let X denote the number of questions she gets correct. Then we have a binomial model with X as our random variable.

For (i), we want $P(X \geq 55)$ so $x = 55$.
For (ii), we want $P(X \geq 60)$ so $x = 60$.
For (iii), we want $P(X \geq 62)$ so $x = 62$.

The approximation applies because she has a 50% chance of getting a correct answer on each question since she guesses, the sample size of $n = 100$ is large, and $x > \frac{n}{2} = 50$ for all three parts of the problem.

(i) Is $\frac{n}{2} < x \leq \frac{n}{2} + 1.1\sqrt{n}$? Yes, because $50 < 55 \leq 61$ so we use the formula.

$$P(X \geq 55)$$
$$\sim \frac{1}{2} - \frac{1}{10}\left[\left(\frac{2 \cdot 55 - 100}{\sqrt{100}}\right)\left\{4.4 - \left(\frac{2 \cdot 55 - 100}{\sqrt{100}}\right)\right\}\right] = .16.$$

(ii) Is $\frac{n}{2} < x \leq \frac{n}{2} + 1.1\sqrt{n}$? Yes, because $50 < 60 \leq 61$ so we use the formula.

$$P(X \geq 60)$$
$$\sim \frac{1}{2} - \frac{1}{10}\left[\left(\frac{2 \cdot 60 - 100}{\sqrt{100}}\right)\left\{4.4 - \left(\frac{2 \cdot 60 - 100}{\sqrt{100}}\right)\right\}\right] = .02.$$

(iii) Is $\frac{n}{2} < x \leq \frac{n}{2} + 1.1\sqrt{n}$? No, because $50 < 62 \nleq 61$. Is $x > \frac{n}{2} + 1.1\sqrt{n}$? Yes, because $62 > 61$. So we use the bound.

$$P(X \geq 62) \leq .01.$$

6.3   Waiting Time Model

This section deals with a model which is similar to the binomial except that the number of draws is not fixed. The random variable of interest in this model is the number of trials needed for an event to occur. For this reason, we call it a waiting time model.

Example 6.12. A box contains eight red and six white chips. Chips are drawn with replacement from the box until three red chips are drawn. Let X be a random variable denoting the total number of chips drawn. Find an expression for (do not evaluate) $P(X = 6)$.

Solution: In order for exactly six chips to be drawn until we get three red chips, we must get two red chips in five draws and then a red chip on the sixth draw.

$$
\begin{aligned}
P(X = 6) \;&=\; P(\text{exactly 6 draws are required to get 3 red chips}) \\[4pt]
&=\; P(\text{draw 2 red chips on the first 5 draws} \\
&\qquad \text{and draw a red chip on the } 6^{th} \text{ draw}) \\[4pt]
&=\; P(\text{2 red chips in 5 draws}) \; P(\text{red chip on the } 6^{th} \text{ draw}) \\[4pt]
&\qquad \text{by independence.}
\end{aligned}
$$

$$
\begin{aligned}
P(X = 6) \;&=\; \left[\frac{5!}{2!3!} \left(\frac{8}{14} \right)^{2} \left(\frac{6}{14} \right)^{3} \right] \left(\frac{8}{14} \right) \\[8pt]
&=\; \frac{5!}{2!3!} \left(\frac{8}{14} \right)^{3} \left(\frac{6}{14} \right)^{3}.
\end{aligned}
$$

We must get three red and three white chips but also the third red chip must be selected on the sixth draw. So the words we are counting are:

$RRWWW \mid R$	$WRWRW \mid R$
$RWRWW \mid R$	$WRWWR \mid R$
$RWWRW \mid R$	$WWRRW \mid R$
$RWWWR \mid R$	$WWRWR \mid R$
$WRRWW \mid R$	$WWWRR \mid R$

The number of words is $5!/(2!3!)$ because we are looking at all possible words with $2R$'s and $3W$'s. For each of these words, we add an R in the last position.

This model is similar to the binomial except that the number of draws is not fixed here. In fact, the number of draws is a random variable. What is fixed in this model is the number of red chips drawn which happens to be the random variable in the binomial model. So the difference between the binomial model and the waiting time model is:

	Number of Draws	Number of Red Chips Drawn
Binomial Model	fixed	random variable
Waiting Time Model	random variable	fixed

The following example uses both the binomial model and the waiting time model.

Example 6.13. A coin has probability $\frac{2}{3}$ of turning up heads.

(i) The coin is tossed until four heads appear. Find an expression for the probability that six tosses are needed.

(ii) The coin is tossed six times. Find the probability of getting four heads.

Solution:

(i) Waiting time model

$$P \text{ (6 tosses are needed to get 4 heads)}$$

$$= \ P \text{ (3 heads in 5 tosses \underline{and} a head on the 6}^{\text{th}} \text{ toss)}$$

$$= \ \left[\frac{5!}{3!2!} \left(\frac{2}{3} \right)^3 \left(\frac{1}{3} \right)^2 \right] \frac{2}{3} = \frac{5!}{3!2!} \left(\frac{2}{3} \right)^4 \left(\frac{1}{3} \right)^2 .$$

(ii) Binomial model

$$P \text{ (4 heads in 6 tosses)} = \frac{6!}{4!2!} \left(\frac{2}{3} \right)^4 \left(\frac{1}{3} \right)^2 .$$

Example 6.14.   (Effect of Home-Away Sequence in a Championship Series.) In most athletic competitions there is a home-field advantage. Familiarity with the teams' own park and local fan support can be important factors in some sports. The final championship series in major league basketball and hockey are seven game series where the first team to win four games wins the series. The team that has the superior regular season record is rewarded by being assigned to play four games at their home field and three games away (at their opponent's home field). In basketball, the home-away sequence is $HHAAAHH$. This means that the team with the better regular season record is scheduled to play two games at home, then three games away, and then two games back home. In hockey the sequence is $HHAAHAH$. Is the probability of winning the series affected by the home-away sequence? For example, suppose we had a seven-game series where the home team gets to play four games at home and the home team has a choice between sequences $HHHHAAA$ or $AAAHHHH$. Is the probability of winning a series affected by which sequence they select?

Before attempting to answer this question regarding the sequence, we introduce some notation. Suppose we have a series between two teams known as the Jazz and the Spurs and suppose the Jazz have a better regular season record than the Spurs so more home games will be played at Jazz Park than at Spurs Park in the championship series.

Let h denote the probability that the Jazz win any game at Jazz Park (h denotes the probability that the Jazz win at home). Let a denote the probability that the Jazz win any game at Spurs Park (a denotes the probability that the Jazz win away).

Then $(1 - h)$ is the probability that the Spurs win any game at Jazz Park. And $(1 - a)$ is the probability that the Spurs win any game at Spurs Park. We assume no ties so each game is either won by the Jazz or the Spurs.

If $h > a$, then the Jazz are more likely to win at Jazz Park than at Spurs Park. Equivalently, the Jazz are more likely to win at home than away. This is called a home-field advantage.

When we multiply an equality by a negative quantity we reverse the direction of the inequality. It follows that if $h > a$, then $(1 - a) > (1 - h)$. The inequality $(1 - a) > (1 - h)$ states that the Spurs are more likely to win at home than away. This is an inequality equivalent to $h > a$ which also states that there is a home-field advantage.

Suppose $h = (1 - a)$. This says that the probability that the Jazz win any game at Jazz Park equals the probability that the Spurs win any game at Spurs Park. Equivalently, both teams are equally likely to win in their home park. This says that both teams are of equal strength. The equation $h = (1 - a)$ can be rewritten $a = (1 - h)$. If $a = (1 - h)$, then the chances that the Jazz win at Spurs park equals the chances that the Spurs win at Jazz Park. In other words, each team has the same probability of winning away. Again, this says that both teams are of equal strength.

Now, we answer the question about whether the order in which the home and away games are played affects the outcome of the series. To simplify computations, instead of a seven-game series, suppose we have a three-game series where the first team to win two games wins the series. Suppose the Jazz have a better regular season record than the Spurs and so two games are scheduled for Jazz Park and one game is scheduled for Spurs Park.

Assume the outcomes from game to game are independent.

Consider the sequence HAH (this means the first and third game are scheduled for Jazz Park and the second game is scheduled for Spurs Park:

Let W denote a win for the Jazz and L a loss for the Jazz.

Then

$$
\begin{aligned}
P(LWW) \;=\; & P(\text{Jazz lose the } 1^{\text{st}} \text{ game at home,}\\
& \text{Jazz win } 2^{\text{nd}} \text{ game away, Jazz win } 3^{\text{rd}} \text{ game at home})\\
\;=\; & (1 - h)ah.
\end{aligned}
$$

Next we calculate the probability that the Jazz win the series for the sequence HAH.

$$
\begin{aligned}
P(\text{Jazz win the series}) \;=\; & P(\text{Jazz win in two games})\\
& + P(\text{Jazz win in 3 games})\\
\;=\; & P(WW) + \{P(WLW) + P(LWW)\}\\
\;=\; & ha + \{h(1 - a)h + (1 - h)ah\}\\
\;=\; & ha + h^2 - ah^2 + ah - ah^2\\
\;=\; & 2ah + h^2 - 2ah^2.
\end{aligned}
$$

(Strictly speaking, we don't have the same waiting time model as described earlier in this section unless $a = h$ because the probability of winning changes from game to game depending on whether the game is at home or away.)

Now consider the sequence AHH:

$P(\text{Jazz win the series})$ must be the same for the sequence AHH as for the sequence HAH because in both cases the Jazz win in two games by

winning one game at home and one game away or the Jazz win in three
games by winning the third game at home and winning exactly one of the
first two games at home.

So, for the sequence AHH,

$$P(\text{Jazz win the series}) = 2ah + h^2 - 2ah^2.$$

Consider the remaining possible sequence which is HHA:

$$
\begin{aligned}
P(\text{Jazz win the series}) &= P(\text{Jazz win in 2 games}) \\
&\quad + P(\text{Jazz win in 3 games}) \\
&= P(WW) + \{P(WLW) + P(LWW)\} \\
&= hh + \{h(1-h)a + (1-h)ha\} \\
&= h^2 + ha - h^2a + ha - h^2a \\
&= h^2 + 2ha - 2h^2a.
\end{aligned}
$$

We see that $P(\text{Jazz win the series})$ is the same regardless of whether the
sequence is HAH or AHH or HHA. It is possible to show that this result
holds more generally. That is, the probability that a team wins the series
does not depend on the home-away sequence.

This example demonstrates that in a championship series as we have
in basketball and hockey where the first team to win four games wins the
series, the probability of a team winning the series does not depend on
the home-away sequence. Baseball has the same home-away sequence as
basketball ($HHAAAHH$) but is different than both hockey and basketball because in baseball the team which is scheduled to play four games
in their home park is determined by which league wins the All-Star game
rather than being based on a superior regular season record. Until recently, the two teams in the baseball world series played regular season
schedules without any common opponents so it would have been difficult
to judge which team had the superior record before the series. Even today, there remains very little overlap in the regular season schedules of the
league champions. But assigning one team four games at home because
their league won the All-Star game is unfair to the other team. A team
has a better chance of winning the series if they have four home games
rather than three home games provided there is a home field advantage
(see Problem 32 at the end of this Chapter). In general, if there is a home
field advantage, a team has a better chance of winning the championship

series when there are more home games for that team. So the team with more home games gains an advantage.

6.4 Problems

1. Find the probability of getting exactly two heads when three fair coins are tossed. Then check your answer by listing both the sample space and the event "exactly two heads."

2. A fair die is tossed three times. Find the probability that the face with six dots turns up at least two times.

3. Marni takes a ten-question true-false exam. Find an expression for the probability that she gets at least nine questions right if:

   (i) She guesses on every question.

   (ii) She knows the answer to four questions, and she guesses on the other six questions.

4. Tina gives the following answer to part (ii) of Chapter 6, Problem 3:
$$\left\{ \frac{(10!)}{(9!)(1!)} \right\} (.5)^5 (.5)^1 + (.5)^6.$$

   She reasons that there are $\left\{ \frac{(10!)}{(9!)(1!)} \right\}$ ways to arrange the $9R$'s and $1W$. Explain the flaw in her argument.

5. In each case state whether the binomial is a reasonable model and explain your answer.

   (i) In a test for the presence of ESP (extra-sensory perception), a subject is asked to identify which of four possible patterns is on a card visible to the experimenter but not to the subject. This is repeated for 100 cards and the number of correct identifications is recorded.

   (ii) A fair coin is tossed until three heads appear. The number of tosses is recorded.

6. In the world series of field hockey the first team to win four games wins the series. Suppose the two teams playing in the series are evenly matched. What is the probability that the series lasts seven games? (Hint: What has to happen in the first six games for the series to last seven games?)

7. Members of a three-person jury, acting independently, each have probability equal to p of making the correct decision. The decision of the majority is final.

   If $p = .4$, does the jury have a better chance of deciding correctly than a single juror? If $p = .7$, does the jury have a better chance of deciding correctly than a single juror?

8. Eight passengers make reservations on a six-seat airplane. The probability that a passenger does not show up for the flight is $\frac{1}{3}$. Find the expected number of people who get bumped due to overbooking.

9. In Example 6.4, suppose I take nine reservations. Let X = number of people who show up. $P(\text{everyone gets a seat}) = P(a \leq X \leq b) = 1 - P(c \leq X \leq d)$. Find the values of a, b, c, d.

10. In Example 6.5 write out the expression for (do not evaluate) $E(\text{gain})$ if they take ten reservations.

11. Refer to Example 6.5. Suppose that when people are bumped they can travel on the next-day flight free of charge, and in addition the company also pays \$4 for overnight accommodations at Motel 4. Write out the expression for (do not evaluate) $E(\text{gain})$ if they take seven reservations.

12. In Example 6.6, verify using a calculator that $p = .903$ satisfies the equation

$$.965 = \frac{8!}{6!2!}p^6(1 - p)^2 + \frac{8!}{7!1!}p^7(1 - p) + p^8.$$

13. Suppose each player who plays a full softball game bats four times. But Brandy bats only three times in a game before being taken out for a defensive replacement. She estimates that in about 60% of the games that she plays she gets at least two hits. What is the probability that she would get at least two hits in a game if she were allowed to stay in a game and bat four times?

14. Roll a balanced die five times. Let B = "at least two 6's" and A = "odd number of 6's." Find $P(A|B)$.

15. In "People vs. Collins," if we assume there are 5,000,000 couples in the L.A. area who could have committed the crime, what is the probability that there is at least one other couple in the area with the six traits?

16. In the modified Walbert model, show that $P(\text{conviction}) = p$ for a two-person jury.

17. In Example 6.10, show that if there are two groups with 15 persons per group the expected number of tests is 6.39.

18. If there are 65 students in class the day of the lottery (see Example 1.11), find the probability that more than one student will select the winning number.

19. Reference [10] contains the following problem:

    "*Sports Illustrated* reports that a high school football team in Bloomington, Indiana lost 21 straight pregame coin flips before finally winning one! We all know the probability of this for a *particular* team is $(1/2)^{21}$, but there are many, many teams. In fact, according to a reliable source, there are approximately 15,000 football teams in the United States, when we consider high school, college, and pro teams."

    The question in this problem relates to the probability of specific numbers of teams losing the last 21 flips. The solution in this reference assumes a binomial model. Discuss whether or not the assumptions for the binomial model are satisfied in this problem.

20. It is known that 20% of all students in the world are left-handed. A class with 20 students has five left-handed and 18 right-handed chairs. Find an expression for (do not evaluate) the probability that each student will have a chair to match his or her handedness.

21. The cashier at Be-Lo Supermarket has a bowl of pennies at his register. He supplies a penny to any customer who owes a penny more than an integer multiple of a dollar and does not want to break a paper bill. Suppose the cashier waits on 20 customers each work period. How many pennies does he need to have in the bowl at the start of his work period so as to be at least 99% certain that he will have enough pennies to satisfy all those customers who are a penny short during his work period? Solve by trial and error.

22. Jennifer has tickets to the fourth and fifth games of the World Series of Women's Lacrosse. In this world series, the first team to win four games wins the series. If the two teams playing in the series are evenly matched, find the expected number of games Jennifer will see.

23. The College has invited 3 rich friends of Margaret Thatcher to a Charter Day picnic. For the purpose of ordering the picnic baskets, it is assumed that the decisions of the guests are unrelated, and based on previous information, it is estimated that each has probability 1/2 of accepting the invitation. How many baskets should be ordered if the College wishes to be 87.5% certain that there will be a basket for each guest who comes? (Use trial and error.)

*24. The College has invited 400 rich friends of Margaret Thatcher to a Charter Day picnic. For the purpose of ordering the picnic baskets, it is assumed that the decisions of the guests are unrelated, and based on previous information, it is estimated that each has probability 1/2 of accepting the invitation. How many baskets should be ordered if the College wishes to be approximately 98% certain that there will be a basket for each guest who comes? (Use trial and error.)

*25. Air Canada and Laker Air offer identical service on two flights leaving North Bay at the same time (meaning the probability of a passenger choosing either is 1/2). Suppose both airlines are competing for the same pool of 400 potential passengers. Air Canada sells tickets to everyone who requests them and the capacity of its plane is 230. Approximate the probability that Air Canada overbooks.

*26. In a 162 game season, find the approximate probability that a team with a 50–50 chance of winning each game will win at least 87 games in a season. (Note: The 2000 New York Yankees won 87 games during the season and went on to win the World Series.)

27. If the probability is .75 that a person will believe a rumor about the transgressions of Bill Clinton, find an expression for (do not evaluate) the probability that the eighth person to hear the rumor will be the fifth person to believe it.

28. A baseball player is facing a pitcher who has a 50% chance of throwing a "ball" and a 50% chance of throwing a "strike" on each pitch. The player's strategy is never to swing at a pitch. Find the probability that he strikes out (i.e., he gets three strikes before he gets four balls).

29. A baseball team has all .300 hitters. Assume players never steal a base, get picked off base, or hit into a double play. Each batter is either out or hits a single. Players on base advance only one base

when there is a single. Find an expression for (do not evaluate) the probability of this team getting one or more runs in an inning.

30. Suppose the Pirates play the Yankees in a five-game series where the first team to win three games wins the series. Suppose the Pirates win any game at home with probability .7 and the Yankees win any game at home with probability .4. Suppose outcomes from game to game are independent. Find the probability that the Pirates win the series in *three* games if:

    (i) The home-away sequence for the Pirates is HHHAA.

    (ii) The home-away sequence for the Pirates is AAHHH.

31. Refer to the previous problem. Find the probability that the Pirates win the series if:

    (i) The home-away sequence for the Pirates is HHHAA.

    (ii) The home-away sequence for the Pirates is AAHHH.

32. Refer to Example 6.14. Suppose there is a three-game series between the Jazz and the Spurs where there is a home-field advantage ($h > a$). Then the Jazz have a better chance of winning the championship series when there are more home games for them. The calculations in this problem provide an example of this.

    (i) If the Jazz play two home games (say the home-away sequence for the Jazz is HAH), find P(Jazz win series) in terms of a and h.

    (ii) If the Jazz play one home game (say the home-away sequence for the Jazz is HAA), find P(Jazz win series) in terms of a and h.

    (iii) Show that the expression in (i) is larger than the expression in (ii) when $h > a$.

33. The 1918 World Series was played between the Boston Red Sox and the Chicago Cubs. To cut down on travel expenses between the two cities, the first three games were scheduled to be played in Chicago and the remaining games (until one team won four games) were to be played in Boston. Assuming that there was a home-field advantage, was this arrangement unfair to one of the teams? Explain. (Note: Harry Frazee, owner of the Red Sox and the man who sold Babe Ruth to the Yankees in 1920, complained that the arrangement was unfair to his team.)

34. In an ESP (extra-sensory perception) experiment, three cards numbered 1, 2, 3 are shuffled and dealt face down in front of a subject. The subject is asked to assign the correct number to each card. If nine subjects are tested and none of them has special powers, find an expression for (do not evaluate) the probability that at least seven subjects will correctly identify all three cards.

35. In an experiment on human behavior, a psychologist asks four men and four women to enter a room and sit at a rectangular table. The table has three seats along each side, and one seat at each end. The end seats are considered to be the dominant positions at the table. The experiment is repeated ten times using new subjects each time. Assuming that seats are randomly chosen each time, find an expression for (do not evaluate) the probability that men occupy both end seats exactly three times.

36. In Problem 8 in Chapter 1, find the probability that the estimate of $P(3$ different faces in 10 simulations$)$ is less than .20.

37. Suppose that it takes at least two votes from a three-member jury to convict a defendant. Suppose that the probability that a juror votes a guilty person innocent is .2, whereas the probability that a juror votes an innocent person guilty is .1. If each juror acts independently and if 65% of the defendants are guilty, find an expression for (do not evaluate) the probability that a defendant is guilty if the jury renders a correct decision.

38. Construct a question which requests either a probability or an expected value relating to the situation described in the following passage. Then describe how you would solve the question you have posed by giving the techniques and explaining how they apply. State your assumptions. Do not write out the formula(s) or do the calculations.

    The following passage is taken from "Summer of '49," by David Halberstam.

    > Monday, May 2, 1949, was the worst day of young Maury Allen's life. He was a local baseball bookie, and he had to pay out. He had been beaten and beaten big. Small baseball pools were not unusual in that era. The idea was quite simple. Other kids, usually friends, would each put down a dime and pick three hitters. If one's chosen

hitters got six hits among them, Allen would have to pay
back sixty cents. Generally such pools were quite prof-
itable: About thirty kids might bet, and it was rare that
more than one or two would beat the system. Allen usually made about eight or ten dollars a week, which was a
nice sideline. It was easy to figure out how his clients
would bet. The Yankee fans would invariably offer up the
three best Yankee hitters. The other kids, the Yankee
haters, would bet on whichever team was in town playing
the Yankees.

On the previous weekend, the Red Sox had been in town,
and there had been a lot of betting, particularly on the
four best Red Sox hitters: Dom Dimaggio, Pesky, Williams,
and Stephens. Maury Allen had a little spiral notebook
and he had dutifully written down the bets: Goldberg–
Pesky, D. Dimaggio, Williams. The Yankees had won the
Friday game behind Vic Raschi, and while Dom Dimaggio
had gotten three hits, Raschi had closed off the middle of
the batting order. There was heavy betting for Saturday
and Sunday as well, but Allen was not worried—on Saturday Eddie Lopat was pitching. Lopat was a bookie's
delight, and a good hitter's nightmare. On Sunday Allie
Reynolds was pitching. The odds could not have looked
better. The Yankees did indeed win on Saturday. But
on Sunday everything came apart for Maury Allen. Dom
Dimaggio opened the game with a single and then Pesky
hit a home run into the right-field seats. That was a
bad omen. Domenic went on to have three hits that day,
Pesky also had three, Williams two, and Bobby Doerr and
Junior Stephens one apiece.

Allen was cooked. Almost everyone who bet with him
had beaten him. He owed fifteen dollars, far more than
he had in reserve. He could not afford to go to school that
Monday, and had to spend the day borrowing from every
member of his family, particularly his brother. That Red
Sox victory, 11–2, did not derail the Yankees from first
place, and it left Boston still in sixth place. But it ended
Maury Allen's career as a bookmaker.

7

Some Simple Statistical Tests

7.1   Introduction

Statistical tests involve testing claims by collecting and analyzing data. There are many problems that are concerned with testing some conjecture or hypothesis. We specify two competing hypotheses. One hypothesis which is called the null hypothesis and denoted by H_0 is the statement that observed results are due to chance. The other hypothesis which is called the alternative hypothesis and denoted by H_1 is the hypothesis we try to establish. We collect data and determine if we should reject H_0 in favor of H_1. The question is whether or not the data are much more consistent with H_1 than H_0 in which case H_0 should be rejected. Statistical hypothesis testing enables us to distinguish chance effects from real ones.

In a jury trial, we would test:

H_0: defendant is not guilty versus H_1: defendant is guilty.

Does the evidence presented to the jury indicate that we should reject H_0 in favor of H_1?

The objective of a test is to determine if the data will discredit the null hypothesis convincingly so as to make the alternative hypothesis seem quite reasonable. If there is some small doubt, then the null hypothesis is not rejected. In the jury trial, one wants to show guilt beyond a reasonable doubt.

The level of significance (or significance level) of a test is the probability of rejecting H_0 when H_0 is true. Rejecting H_0 when H_0 is true is an error. The more serious the error, the smaller should be the significance level. In the jury trial example, the error would be finding the defendant guilty when she is not guilty. We would like the probability of this to be small.

★ 7.2   Binomial Tests

Consider the binomial model: $P(X = x) = \frac{n!}{x!(n-x)!} \cdot p^x(1 - p)^{n-x}$, $x = 0, 1, 2, 3, \ldots, n$.

Suppose we want to test $H_0 : p = p_0$ versus $H_1 : p = p_1$ where p_0 and p_1 are specific values of p. We need a principle which will determine the form of the test.

The principle is as follows:

Reject H_0 when the ratio

$$L = \left[\frac{P(X = x \text{ when } H_1 \text{ is true})}{P(X = x \text{ when } H_0 \text{ is true})} \right] \text{ is large.}$$

This makes sense intuitively because we would want to reject H_0 when our data are more consistent with H_1 than H_0.

Example 7.1. A box has red chips and white chips. Suppose we know that the ratio of red chips to white chips in the box is either 1 or 2. That is, either there are the same number of reds and whites in the box or there are twice as many reds as whites in the box. Let p denote the proportion of red chips in the box.

We wish to test H_0: $p = \frac{1}{2}$ versus H_1: $p = \frac{2}{3}$.

Suppose the experiment consists of drawing three chips with replacement from the box. On the basis of the data we will decide whether or not to reject H_0.

Let X denote the number of red chips drawn. When either H_0 or H_1 is true, we have a binomial model.

Suppose we calculate the ratio L for an arbitrary value of x, say $x = 1$.

$$P(X = 1 \text{ when } H_1 \text{ is true}) = \frac{3!}{1!2!} \left(\frac{2}{3}\right)^1 \left(\frac{1}{3}\right)^2 = \frac{6}{27};$$

$$P(X = 1 \text{ when } H_0 \text{ is true}) = \frac{3!}{1!2!} \left(\frac{1}{2}\right)^1 \left(\frac{1}{2}\right)^2 = \frac{3}{8};$$

$$L = \frac{\frac{6}{27}}{\frac{3}{8}} = \frac{16}{27}.$$

Next, we calculate the ratio

$$L = P(X = x \text{ when } H_1 \text{ is true})/P(X = x \text{ when } H_0 \text{ is true})$$

for all possible values of x and put the results in table form.

x	$P(X = x \text{ when } H_0 \text{ is true})$	$P(X = x \text{ when } H_1 \text{ is true})$	L
0	$\frac{1}{8}$	$\frac{1}{27}$	$\frac{8}{27}$
1	$\frac{3}{8}$	$\frac{6}{27}$	$\frac{16}{27}$
2	$\frac{3}{8}$	$\frac{12}{27}$	$\frac{32}{27}$
3	$\frac{1}{8}$	$\frac{8}{27}$	$\frac{64}{27}$

According to our principle for determining the form of the test, we should reject H_0 when L is large. The table indicates that L increases as x increases. So we should reject H_0 when L is large or equivalently when X is large. This makes sense intuitively since a large value of X is more consistent with H_1 than H_0 and we would want to reject H_0 when there are a large number of red chips drawn. Another way to indicate that X is "large" is $X \geq c$ where c is constant.

So we should reject H_0 when $X \geq c$. What is the value of c?

Should we reject H_0 when $X \geq 2$? Should we reject H_0 when $X \geq 3$?

The principle only gives us the form of the answer. It doesn't give the value of c.

We would like $P(\text{reject } H_0 \text{ erroneously})$ to be small. In the jury trial example, we wouldn't want an innocent defendant to be found guilty very often.

In our example,

$$
\begin{aligned}
P(\text{reject } H_0 \text{ erroneously}) &= P(\text{reject } H_0 \text{ when } H_0 \text{ is true}) \\
&= P(X \geq c \text{ when } H_0 \text{ is true}).
\end{aligned}
$$

We calculate $P(X \geq c$ when H_0 is true$)$ for different values of c from the table:

$$P(X \geq 3 \text{ when } H_0 \text{ is true}) \quad = \quad \frac{1}{8} = .125.$$

$$P(X \geq 2 \text{ when } H_0 \text{ is true}) \quad = \quad \frac{3}{8} + \frac{1}{8} = .5.$$

$$P(X \geq 1 \text{ when } H_0 \text{ is true}) \quad = \quad \frac{3}{8} + \frac{3}{8} + \frac{1}{8} = .875.$$

So, if we reject H_0 when $X \geq 2$ there is a 50% chance of rejecting H_0 erroneously. It might be better to reject H_0 when $X \geq 3$ and have a 12.5% chance of rejecting H_0 erroneously. In general, selecting a specific value for c depends on the particular situation and the consequences of making an erroneous decision. If we make the chances of erroneously rejecting H_0 too small, we would frequently fail to reject H_0 when H_1 was true. In the jury example, guilty defendants would rarely be found guilty. In an actual problem, rather than specifying a value for c and then calculating P(rejecting H_0 erroneously), the significance level is specified first and the value of c is determined from that. In our example, if we specified a significance level of .125, then the value of c would be 3. Of course, in examples such as this one there will only be a few significance levels possible since X can assume only a few values.

Rather than a specific value for p when H_1 is true, it is usually more realistic to have an interval for p when H_1 is true. In our example, we might want to test H_0: $p = \frac{1}{2}$ versus H_1: $p > \frac{1}{2}$. Recall that $p = P(\text{red})$. So H_0 corresponds to a box with the same number of reds and whites and H_1 corresponds to a box with more reds than whites. We could make up another table for any specific $p > \frac{1}{2}$ and even though the values of L would change, they would still increase as x increases meaning the criteria for rejecting H_0 would be the same. So, H_0: $p = \frac{1}{2}$ versus H_1: $p = \frac{2}{3}$ and H_0: $p = \frac{1}{2}$ versus H_1: $p > \frac{1}{2}$ lead to the same decision rule (see Exercise 1 in Chapter 7).

The following example is based on an actual case. The numbers have been changed here to make the computations simpler. The example which includes the actual values can be found in Problem 9 in Chapter 7.

Example 7.2. In elections, candidates are supposed to be listed on the ballot in randomly chosen order. Nicholas Caputo, the clerk of Essex County, NJ, had assigned first position on the ballot forms to a Democrat

11 times out of 12. He was taken to court on the grounds that such an occurrence was not consistent with random choices. Use a significance level of .02 to test whether the data are not consistent with a hypothesis of random selection.

Let $p = P(\text{Democrat is listed first on the ballot})$.

If the choice is random, then $p = \frac{1}{2}$. This will be our null hypothesis. It says that Democrats and Republicans are equally likely to be listed first on the ballot. The alternative hypothesis that we are trying to establish is that Democrats were more likely to be assigned first position on the ballot by Nicholas Caputo. So the alternative hypothesis is $p > \frac{1}{2}$. Let X denote the number of times the Democrat is listed first on the ballot. In binomial tests, the random variable, X, is the number of times the event whose probability is stated by p occurs.

We test H_0: $p = \frac{1}{2}$ versus H_1: $p > \frac{1}{2}$. A large value of X is more consistent with H_1 than H_0 and we would want to reject H_0 when $X \geq c$. The principle to determine the form of the test which involves calculating the ratio L would also lead to the criteria of rejecting H_0 when X is large. The reason for calculating the ratio L for a test that assumes a population is that it provides the form of the best test available. In the examples given here it is not difficult to determine which summary statistic to use based on the data and to state the form of the test. But in other more complicated situations, the decision criteria will not be known until after the ratio L is determined.

Next, we calculate c. The value of c is determined by the significance level.

$$P(X \geq 12 \text{ when } H_0 \text{ is true}) = \left(\frac{1}{2}\right)^{12} = .0002.$$

$$P(X \geq 11 \text{ when } H_0 \text{ is true}) = \frac{12!}{11!1!}\left(\frac{1}{2}\right)^{11}\left(\frac{1}{2}\right)^{1} + \left(\frac{1}{2}\right)^{12} = .003.$$

$$P(X \geq 10 \text{ when } H_0 \text{ is true}) = \frac{12!}{10!2!}\left(\frac{1}{2}\right)^{10}\left(\frac{1}{2}\right)^{2}$$

$$+ \frac{12!}{11!1!}\left(\frac{1}{2}\right)^{11}\left(\frac{1}{2}\right)^{1} + \left(\frac{1}{2}\right)^{12} = .02.$$

Since we have selected a significance level of .02, the value of c is 10 and we should reject H_0 when $X \geq 10$. Since $X = 11$ from the data, we reject H_0 and conclude that Democrats were more likely to be assigned first position on the ballot. In the actual case, the New Jersey Supreme

Court told Democrat Caputo to cease his practices according to the *New York Times* of August 13, 1985.

The following example shows how the binomial model can be used to test a hypothesis about two populations when the data are paired.

Example 7.3. The article, "Sex and Race Discrimination in the New Car Showroom: A Fact or Myth," in J. Consumer Affairs, 1977, reported results in which individuals of different race and sex visit car dealerships requesting the best possible deal on a certain car. The following are representative data of amounts paid (in dollars):

Dealer	1	2	3	4	5
Black Female	4459	4320	4268	4585	4736
White Male	4348	4385	4231	4516	4550

Dealer	6	7	8	9
Black Female	4262	4440	4498	4823
White Male	4203	4285	4408	4570

Should we conclude that white males have a greater chance of getting a better deal than black females, using a significance level of .02?

Solution: Let $p = P$(white males get a better deal than black females).
    We want to test

$$H_0: p = \tfrac{1}{2}\text{(no discrimination)}$$

versus

$$H_1: p > \tfrac{1}{2}\text{(discrimination against black females)}.$$

So if BF stands for "black female gets a better deal" and WM stands for "white male gets a better deal," imagine that we have a box which has an equal number of BF and WM chips when H_0 is true but more WM chips than BF chips when H_1 is true. We draw nine chips with replacement from the box, each chip corresponding to a dealer.

    Let X denote the number of times a white male gets a better deal than a black female.

    The principle to determine the form of the test would lead to the decision rule of rejection of H_0 when X is large. This is also reasonable on intuitive grounds since a large value of X is more consistent with H_1 than H_0.

So we reject H_0 when $X \geq c$.

We consider some possible values for c and the corresponding significance levels.

When $c = 9$,

$$P(X \geq 9 \text{ when } H_0 \text{ is true}) = \left(\frac{1}{2}\right)^9 = .002.$$

When $c = 8$,

$$P(X \geq 8 \text{ when } H_0 \text{ is true}) = \frac{9!}{8!1!}\left(\frac{1}{2}\right)^8\left(\frac{1}{2}\right)^1 + \left(\frac{1}{2}\right)^9 = .02.$$

When $c = 7$,

$$P(X \geq 7 \text{ when } H_0 \text{ is true})$$

$$= \frac{9!}{7!2!}\left(\frac{1}{2}\right)^7\left(\frac{1}{2}\right)^2 + \frac{9!}{8!1!}\left(\frac{1}{2}\right)^8\left(\frac{1}{2}\right)^1 + \left(\frac{1}{2}\right)^9 = .09.$$

Some possible significance levels are .002, .02, .09.

We chose a significance level of .02. This means we reject H_0 when $X \geq 8$.

According to the data, $x = 8$ because for eight of the dealers (all but dealer 2) the white male got a better deal than the black female. Therefore we reject H_0 and conclude that white males have a greater chance of getting a better deal than black females. For any test, the significance level chosen is crucial to the decision reached.

The following example shows how the binomial model can be used to test for trend.

Example 7.4. The following are average grade point averages for spring semester graduates from 1977 to 1990:

1977	2.76		1984	2.88
1978	2.74		1985	2.80
1979	2.73		1986	2.86
1980	2.72		1987	2.88
1981	2.80		1988	2.82
1982	2.80		1989	2.87
1983	2.84		1990	2.88

Is there an increasing trend if the significance level is .063?

Solution: To answer this question, we split the data in half (if there were an odd number of points we would discard the middle value). We then compare corresponding pairs of average grade point averages. We compare the first reading in the first half of the data with the first reading in the second half of the data. That is, we compare the 1977 and 1984 grade point averages. Then we compare the corresponding second readings, etc. In this way, the comparisons are between two years which are always the same distance apart.

If there is no trend, we are just as likely to get an increase (I) as a decrease (D) for each pair. Let $p = P(I)$. Then we are testing

$$H_0: p = .5(\text{no trend}) \text{ versus } H_1: p > .5(\text{increasing trend}).$$

Let X denote the number of increasing pairs (I's). We reject H_0 when X is large. So we reject H_0 when $X \geq c$. We selected a significance level of .063.

Then $c = 6$ because

$$P(X \geq 6 \text{ when } H_0 \text{ is true}) = \frac{7!}{6!1!} \left(\frac{1}{2}\right)^6 \left(\frac{1}{2}\right)^1 + \left(\frac{1}{2}\right)^7 = .063.$$

From the data we see that $x = 7$. That is, for all seven years there is an increase. The decision rule is to reject H_0 when $X \geq 6$ but $x = 7$, so we reject H_0 and conclude that there is an increasing trend. The data indicate grade inflation.

Had the value of x from the data set been five rather than seven, we would have concluded that there was not enough evidence to reject the null hypothesis.

For these examples using small samples, we are very limited in possible choices of levels of significance. It is common practice in actual studies to select a significance level near .05.

⋆ 7.3 Tukey's Quick Test

The tests described in Section 7.2 suppose a population. They were developed assuming a sample is drawn from that population. Inferences were then made about the population based on the information in the sample. The example about sex and race discrimination in the new-car showroom is of this type. The conclusion applies generally rather than to the nine dealers in the sample.

There are other types of tests which are developed under the sole assumption of random assignment. These tests refer only to the subjects in the particular study at hand. For randomization tests, the conclusion cannot legitimately be extended to subjects outside the study because there is no assumption that the sample was selected from a population according to some specific sampling plan. In some situations, subjects are selected because they are readily available rather than randomly selected. A randomization test will determine whether or not a certain pattern in the observations is likely to have arisen by chance alone. The significance level is found by comparing the test statistic with the distribution obtained by randomly re-ordering the data in some form. The principle which involves calculating the ratio L does not apply for these tests. An example of a randomization test follows.

Tukey's Quick Test

For this test, the null hypothesis is that two groups are identical. The alternative hypothesis is that one group tends to have larger values. Suppose we were to list the values from the combined sample from the two groups in order of increasing magnitude. We call the groups A and B.

We test H_0: A's and B's are equal versus H_1: A's tend to be larger than B's.

If H_0 is true, all possible arrangements of A's and B's are equally likely.

An arrangement of the combined samples like BABBBAAAA might lead us to reject H_0 because most A's are larger than most B's. An arrangement like ABABAABAB would lead us not to reject H_0. We will measure this tendency with the random variable

$$X = (\text{number of } A\text{'s to the right of the largest } B)$$
$$+ (\text{number of } B\text{'s to the left of the smallest } A).$$

The sequence must begin at the left end with a B and end at the right end with an A or we set $X = 0$.

Some examples are

$$\underline{B}ABBB\underline{AAAA} \quad X = 1 + 4 = 5;$$
$$ABBBBAAAA \quad X = 0;$$
$$\underline{B}ABABAAB\underline{A} \quad X = 1 + 1 = 2.$$

A large value of X is more consistent with H_1 than H_0 and so we should reject H_0 if $X \geq c$.

Example 7.5. Suppose there are 5 A's and 3 B's. A total of $\frac{8!}{5!3!} = 56$ words are possible and if H_0 is true they are equally likely. List all words where $X \geq 6$:

$$BBBAAAAA \quad X = 8;$$

$$BBABAAAA \quad X = 6.$$

These are the only two cases where $X \geq 6$. Thus,

$$\text{level of significance} = P(X \geq 6 \text{ when } H_0 \text{ is true}) = \frac{2}{56} = .035.$$

So we would reject H_0 if $X \geq 6$ to get a test with significance level of .035 for 5 A's and 3 B's.

Example 7.6. Suppose there are 9 A's and 5 B's. Then there are a total of $\frac{14!}{9!5!} = 2002$ equally likely words when H_0 is true. Rather than listing words because there are many possibilities here, let's count words in a more orderly fashion.

Suppose we want $P(X = 7 \text{ when } H_0 \text{ is true})$.

Then $BA \cdots BAAAAAA$ is part of a word where $x = 7$. We have 6 A's to the right of the largest B and 1 B to the left of the smallest A. We have used up 7 A's and 2 B's for this part of the word. There are $9 - 7 = 2 A$'s and $5 - 2 = 3 B$'s which have not been used. These remaining letters form $\frac{5!}{2!3!}$ possible words which can fill in the blank. We count the total number of words with $X = 7$ using this technique.

Pattern	Letters Used	Letters Unused	No. Words to Fill Blank
$BA \cdots BAAAAAA$	$7A$'s $2B$'s	$2A$'s $3B$'s	$\dfrac{5!}{2!3!} = 10$
$BBA \cdots BAAAAA$	$6A$'s $3B$'s	$3A$'s $2B$'s	$\dfrac{5!}{3!2!} = 10$
$BBBA \cdots BAAAA$	$5A$'s $4B$'s	$4A$'s $1B$	$\dfrac{5!}{4!1!} = 5$
$BBBBA \cdots BAAA$	$4A$'s $5B$'s	$5A$'s $0B$'s	$\dfrac{5!}{5!0!} = 1$

So the number of words with $X = 7$ is

$$\frac{5!}{2!3!} + \frac{5!}{3!2!} + \frac{5!}{4!1!} + \frac{5!}{5!0!} = 10 + 10 + 5 + 1 = 26.$$

$P(X = 7 \text{ when } H_0 \text{ is true}) = \frac{26}{2002} = .013.$

Next, suppose we want $P(X \geq 6$ when H_0 is true$)$. The same method can be used to find all the probabilities $P(X = 6$ when H_0 is true$)$, $P(X = 7$ when H_0 is true$) \dots P(X = 12$ when H_0 is true$)$. It is not possible for $X = 13$ when there are 9 A's and 5 B's (see Problem 13 in Chapter 7). And the only way for $X = 14$ is when we have the arrangement $BBBBBAAAAAAAAA$ so that $P(X = 14$ when H_0 is true$) = \frac{1}{2002}$.

Carrying out the computations gives

$$
\begin{aligned}
P(X \geq 6 \text{ when } H_0 \text{ is true}) \quad = \quad & P(X = 6 \text{ when } H_0 \text{ is true}) \\
& + P(X = 7 \text{ when } H_0 \text{ is true}) \\
& + \cdots + P(X = 12 \text{ when } H_0 \text{ is true}) \\
& + P(X = 14 \text{ when } H_0 \text{ is true}) = .05.
\end{aligned}
$$

So if there are 9 A's and 5 B's, a test with significance level of .05 is to reject H_0 when $X \geq 6$.

There are tables available for this test which specify that when there are 9 A's and 5 B's and the significance level is .05, H_0 should be rejected when $X \geq 6$. This example shows how these tables were constructed.

The next example shows us how to apply Tukey's Quick Test to a data set.

Example 7.7. Nine male students and five female students were asked to give the number of hours of sleep they got on the previous night. The results (in hours) are as follows:

    males : 6.5, 5, 6, 8, 7.5, 3, 3.5, 8.5, 2;   females : 7, 1, 4, 4.5, 5.

Using a significance level of .05, test the null hypothesis that males and females got an equal amount of sleep versus the alternative hypothesis that males got more sleep.

Solution: Denote the male students by A and female students by B. Look at the combined ordered sample from the two groups along with the corresponding group letter.

1	2	3	3.5	4	4.5	5	5	6	6.5	7	7.5	8	8.5
B	A	A	A	B	B	B	A	A	A	B	A	A	A

From Example 7.6, for a significance level of .05, we should reject the null hypothesis when $X \geq 6$. Examination of the above data set reveals that

$X = 3 + 1 = 4$. So there is not enough evidence to conclude that the males got more sleep.

This test was developed by John Tukey and does not involve heavy computations so it is a quick and simple test to carry out. Tukey's Quick Test is presented here because it is easy to understand rather than because it has good properties.

Note: In Example 7.7, there is one A value at 5 and one B value at 5. But no matter which of the 5's we list first, the value of x is unaffected. Sometimes, the tied values are relevant. For details, see reference [40] in the Bibliography.

⋆ 7.4   Problems

1. Show that in Example 7.1, if we test H_0: $p = \frac{1}{2}$ versus H_1: $p = \frac{3}{4}$ we get the same decision criteria as in testing H_0: $p = \frac{1}{2}$ versus H_1: $p = \frac{2}{3}$. Do this by calculating the ratio L for all possible values of x in the former test.

2. Let X denote the number of heads in two tosses of a coin. Let p denote P(head). We wish to test H_0: $p = \frac{1}{2}$(fair coin) versus H_1: $p = \frac{1}{4}$(tails three times as likely as heads).

   Make up a table giving all possible values of x along with the corresponding values of L. Determine the form of the test. If the level of significance is $\frac{1}{4}$, should we reject H_0 if $x = 1$? Explain.

3. In order to decide whether pollution-control devices decrease gas mileage, the following test is conducted. Five gallons of gas are put in each of eight new cars and they are driven at a constant speed until the gas runs out. Then all pollution control devices are disconnected and the experiments repeated. Seven of the eight cars have increased gas mileage in the second trial. Is this evidence that pollution-control devices decrease gas mileage? Choose a significance level of .035.

4. Seven Bennington students went on a diet in an attempt to lose weight. Suppose that six of the seven students lost weight. Is the diet an effective means of losing weight? Choose a significance level of .063.

5. The manager of a women's softball team was concerned about too many of his players getting thrown out at second base on hits that he felt should have been doubles. So he carried out an experiment which involved timing ten of his players to see how long it took each player to run from home plate to second base. A player has some freedom in the path she takes from home plate to second base as long as she stays within the baselines and touches first base. The manager had each player round first base using two different routes: Taking a very wide angle around first base and then following nearly a straight line from first base to second base; taking a narrow angle around first base so that the path from home to second base is approximately a semi-circle. The manager found that for nine of the ten players, the wide-angle path was faster. Would you conclude that the wide-angle path enables runners to reach second base quicker? Choose a significance level of .055.

6. The following are average grade point averages for students graduating at the end of the spring term with BA/BS in the indicated years from 1981-1990. Does a decreasing trend exist in the sequence? That is, has there been grade deflation in the 80's?

	1981	1982	1983	1984	1985
QPA	2.952	2.9251	2.943	2.918	2.899

	1986	1987	1988	1989	1990
QPA	2.899	2.9250	2.871	2.894	2.936

Choose a significance level of .031.

7. The following are divorce rates (rate per 1000 population) in the U.S. in the indicated years according to the 1984 World Almanac. Would you conclude from these data that there has been an increase in the divorce rate?

1915	1920	1925	1930	1935	1940	1945
1.0	1.6	1.5	1.6	1.7	2.0	3.5

1950	1955	1960	1965	1970	1975
2.6	2.3	2.2	2.5	3.5	4.8

Choose a significance level of .016.

8. Mathematician/magician, Persi Diaconis, claims that in the long run, he can toss a fair coin and make it come up heads more than 50% of the time. If Persi tosses a fair coin 100 times and it comes up heads 55 times, would you buy Persi's claim? Use a significance level of .05.

9. In elections, candidates are supposed to be listed on the ballot in randomly chosen order. Nicholas Caputo, the clerk of Essex County, NJ, had assigned first position on the ballot forms to a Democrat 40 times out of 41. He was taken to court on the grounds that such an occurrence was not consistent with random choices. Use a significance level of .03 to test whether the data are not consistent with the hypothesis of random selection. Use the approximation to binomial sums.

10. If there are 7 A's, 5 B's, find an expression for (do not evaluate) $P(X = 6$ when H_0 is true) where X is the random variable in Tukey's Quick Test.

11. If there are 7 A's, 5 B's, find $P(X \geq 9$ when H_0 is true) where X is the random variable in Tukey's Quick Test.

12. If there are 7 A's, 5 B's, find an expression for (do not evaluate) $P(X = 3$ when H_0 is true) where X is the random variable in Tukey's Quick Test.

13. Show that $X = 13$ is not possible when there are 9 A's and 5 B's.

14. Five male students and four female students were asked to give the number of hours of television they watched on the previous night. The results (in hours) are as follows:

    males : 3.5, 4.5, 2, 1, 3;   females : 2.5, 0, 0, 1.5.

Using a significance level of .063, test the null hypothesis that males and females watched the same amount of television versus the alternative hypothesis that males watched more television.

Short Answers to Selected Exercises

Chapter 1

2) (i) Yes.

(ii) 2/9

3) .25

4) .5

6) (i) .333

(ii) 0

7) 5/9

12) .333

13) 2/3

14) 2/3

15) (i) .75

20) .667

Chapter 2

1) 5/9

4) (i) .48

(ii) .52

5) 3/4

8) .0127

9) .500477

Twenty-three people are needed to have better than a 50-50 chance that at least two of the people have the same birthday whereas 253 people are needed to have better than a 50-50 chance that at least one of the people has a birthday on a specific date.

11) .618

14) (i) .35

(ii) .65

(iii) .993

16) (i) .0926

(ii) .9074

(iii) .5981

18) (ii) No. Al could sit in a corner desk in the first row.

19) (i) 1/3

(ii) 1/3

(iii) Yes, because A and B are independent events.

24) .802

25) .000379

28) .7874

29) .0313

30) .936

35) .001

Chapter 3

1) (i) .52

   (ii) .50

4) (i) 5/9

   (ii) 5/9

6) .75

13) .58

14) .15

15) 1/30

16) .42

21) .4

22) .921

23) .14

25) .913

26) $P(B_1|T) = 6/11,$
    $P(B_2|T) = 3/11,$
    $P(B_3|T) = 2/11.$

31) (i) 1/3

    (ii) 8/9

32) .25

38)

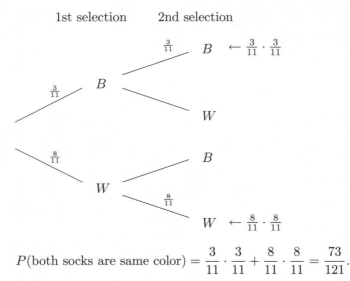

1st selection   2nd selection

$$P(\text{both socks are same color}) = \frac{3}{11} \cdot \frac{3}{11} + \frac{8}{11} \cdot \frac{8}{11} = \frac{73}{121}.$$

41) **(i)** $1/3$

(ii) $3/7$

44) **(i)** $.032967$

(ii) Yes.

(iii) THE SHADOW KNOWS. This was the answer to the question, "Who knows what evil lurks in the hearts of men?" in the popular radio show of the 1940's. A movie based on this show and starring Alec Baldwin as The Shadow was released by Universal Pictures in 1994. The Shadow used a telepathic ability to cloud the minds of evildoers so that they couldn't see him.

Chapter 4

1) **(ii)** $2/3$

4)

x	$P(X = x)$
1	$\frac{1}{6}$
2	$\frac{1}{2}$
3	$\frac{1}{6}$
4	$\frac{1}{6}$

5) 2.5

7) 2.1

8) **(i)** 1

   (ii) 1

10) 15/8

12) $E(X) = 33.667; E(Y) = 25$

19) Entries are $P(X = x, Y = y)$:

x

		0	1	2
	0	$\frac{1}{3}$	0	$\frac{1}{6}$
y	1	0	$\frac{1}{3}$	0
	2	$\frac{1}{6}$	0	0

20) **(ii)** 1/3

23) Entries are $P(X = x, Y = y)$:

x

		1	2
	0	.3	.18
y	1	.2	.24
	2	0	.08

Chapter 5

1) **(i)** 6

3) .000388

7) .135

8) $\dfrac{3!}{2!1!} \left(\dfrac{9}{20} \cdot \dfrac{8}{19} \cdot \dfrac{11}{18} \right) + \dfrac{3!}{3!0!} \left(\dfrac{9}{20} \cdot \dfrac{8}{19} \cdot \dfrac{7}{18} \right)$

9) $\dfrac{5!}{2!3!} \left(\dfrac{15}{35} \cdot \dfrac{14}{34} \cdot \dfrac{20}{33} \cdot \dfrac{19}{32} \cdot \dfrac{18}{31} \right) + \dfrac{5!}{3!2!} \left(\dfrac{15}{35} \cdot \dfrac{14}{34} \cdot \dfrac{13}{33} \cdot \dfrac{20}{32} \cdot \dfrac{19}{31} \right)$

10) $\dfrac{3!}{2!1!}\left(\dfrac{5}{10}\cdot\dfrac{4}{9}\cdot\dfrac{1}{8}\right)$

16) (ii) 23

17) 4

19) 10/19

Chapter 6

1) 3/8

2) 2/27

6) 5/16

8) .234

9) $a = 0; \quad b = 6; \quad c = 7; \quad d = 9$

13) .78

15) .194

20) $\dfrac{20!}{2!18!}(.2)^2(.8)^{18} + \dfrac{20!}{3!17!}(.2)^3(.8)^{17} + \dfrac{20!}{4!16!}(.2)^4(.8)^{16} + \dfrac{20!}{5!15!}(.2)^5(.8)^{15}$

23) Order 2 baskets.

24) 219

27) $\dfrac{7!}{4!3!}\left(\dfrac{3}{4}\right)^5\left(\dfrac{1}{4}\right)^3$

28) 21/32

29) $1 - \left[(.7)^3 + \dfrac{3!}{2!1!}(.7)^3(.3)^1 + \dfrac{4!}{2!2!}(.7)^3(.3)^2 + \dfrac{5!}{2!3!}(.7)^3(.3)^3\right]$

34) $\dfrac{9!}{7!2!}\left(\dfrac{1}{6}\right)^7\left(\dfrac{5}{6}\right)^2 + \dfrac{9!}{8!1!}\left(\dfrac{1}{6}\right)^8\left(\dfrac{5}{6}\right)^1 + \left(\dfrac{1}{6}\right)^9$

36) .00406

Chapter 7

5) Yes.

11) $\dfrac{1}{198}$

12) $\left(\dfrac{7!}{4!3!} + \dfrac{7!}{5!2!} \right) / \left(\dfrac{12!}{7!5!} \right)$

Bibliography

[1] Albert, James H. "College Students' Conceptions of Probability." *The American Statistician* 57 (2003), pp. 37–45.

[2] Andel, Jiri. *Mathematics of Chance*. New York: John Wiley and Sons, 2001.

[3] Austin, J. D. "Overbooking Airline Flights." *The Mathematics Teacher* 75 (1982), pp. 221-223.

[4] Bassett, G. W. and W. J. Hurley. "The Effects of Alternative HOME-AWAY sequences in a Best-of-Seven Playoff Series." *The American Statistician* 52 (1998), pp. 51–53.

[5] Basta, L. L. *Life and Death on Your Own Terms*. Amherst, New York: Prometheus Books, 2001.

[6] Bennett, W. R., Jr. "How Artificial is Intelligence?" *American Scientist*, 65 (1977), pp. 694–702.

[7] Blackwell, David. *Basic Statistics*. New York: McGraw-Hill Book Company, 1969.

[8] Bolstad, W. M., Lyn A. Hunt, and Judith L. McWhirter. "Sex, Drugs and Rock & Roll Survey in a First-Year Service Course in Statistics." *The American Statistician* 55 (2001), pp. 145–149.

[9] Borack, Jules I. "A Technique for Estimating the Probability of Detecting a Nongaming Drug User." *The American Statistician* 51 (1997), pp. 134–136.

[10] Carr, Wendell E. *Statistical Problem Solving*. New York: Marcel Dekker, Inc., 1992.

[11] Chernoff, Herman. "An Analysis of the Massachusetts Numbers Game." *M.I.T. Technical Report #23* (1980), pp. 1–39.

[12] Conover, W. J. *Practical Nonparametric Statistics.* New York: John Wiley & Sons, 1971.

[13] Dittmar, Joseph J. *Baseball Records Registry.* Jefferson, NC: McFarland & Company, Inc., 1997.

[14] Downs, Tom, Dennis C. Gilliland, and Leo Katz. "Probability in a Contested Election." *The American Statistician* 32 (1978), pp. 122–125.

[15] Everitt, Brian S. *Chance Rules.* New York: Springer-Verlag, 1999.

[16] Fairley, William B. and Frederick Mosteller. "A Conversation about Collins." In *Statistics and Public Policy*, edited by William B. Fairley and Frederick Mosteller. Reading, MA: Addison-Wesley, 1977, pp. 369–379.

[17] Falk, Ruma. *Understanding Probability and Statistics.* Wellesley, MA: A. K. Peters, Ltd., 1993.

[18] Field, David A. "Investigating Mathematical Models." *American Mathematical Monthly* 85 (1978), pp. 196–197.

[19] Finkelstein, Michael O. and Bruce Levin. *Statistics for Lawyers.* New York: Springer-Verlag, 1990.

[20] Fox, James A. and Paul E. Tracy. *Randomized Response: A Method for Sensitive Surveys.* Beverly Hills: Sage Publications, Inc., 1986.

[21] Friedlander, R. J. "Ol Abner Has Done it Again," *American Mathematical Monthly* 99 (1992), p. 16.

[22] Gardner, Geoffrey Y. "Computer Identification of Bullets." In *Carnahan Conference on Crime Countermeasures.* Lexington, KY: ORES Publications, 1977, pp. 149–166.

[23] Ghahramani, Saeed. *Fundamentals of Probability.* Upper Saddle River, NJ: Prentice Hall, 2000.

[24] Goldberg, Kenneth P. "Using Technology to Understand the Jury Decision-making Process." *The Mathematics Teacher* 87 (1994), pp. 110–114.

[25] Hader, R. J. "Random Roommate Pairing of Negro and White Students." *The American Statistician* 21 (1967), pp. 24–26.

[26] Halberstam, David. *Summer of '49.* New York: Avon Books, 1989.

[27] Hill, Ted. "The First Digit Phenomenon." *American Scientist* 86 (1998), pp. 358–363.

[28] Hurley, W. J. "What Sort of Tournament Should the World Series Be?" *Chance* 6 (1993), pp. 31–33.

[29] Kalbfleisch, J.G. *Probability and Statistical Inference, Volume 1: Probability.* New York: Springer-Verlag, 1985.

[30] Konold, Clifford. "Teaching Probability through Modeling Real Problems." *The Mathematics Teacher* 87 (1994), pp. 232–235.

[31] Kotz, Samuel and Donna F. Stroup. *Educated Guessing: How to Cope in an Uncertain World.* New York: Marcel Dekker, Inc., 1983.

[32] Larsen, Richard J., and Morris L. Marx. *An Introduction to Mathematical Statistics and its Applications.* Englewood Cliffs, NJ: Prentice Hall, 1986.

[33] Larsen, Richard J., and Morris L. Marx. *An Introduction to Probability and its Applications.* Englewood Cliffs, NJ: Prentice-Hall, Inc., 1985.

[34] LeValliant, Ted and Marcel Theroux. *WHAT'S THE VERDICT?* New York: Sterling, 1991.

[35] Levy, Paul S., Jason Hsia, Borko Jovanovic, and Douglas Passaro. "Sampling Mailrooms for Presence of Anthrax Spores: A Curious Property of the Hypergeometric Distribution under an Unusual Hypothesis Testing Scenario." *Chance* 15 (2002), pp. 19–21.

[36] Ling, Robert F. "Just Say No to Binomial (and other Discrete Distributions) Tables." *The American Statistician* 46 (1992), pp. 53–54.

[37] Litwiller, Bonnie H. and David R. Duncan. "Matching Garage-Door Openers." *The Mathematics Teacher* 85 (1992), pp. 217–219.

[38] May, E. Lee. "Are Seven-Game Baseball Playoffs Fairer?" *The Mathematics Teacher* 85 (1992), pp. 528–531.

[39] Mosteller, Frederick. *Fifty Challenging Problems in Probability with Solutions*. Reading, MA: Addison-Wesley, 1965.

[40] Neave, Henry R. and Peter L. Worthington. *Distribution-Free Tests*. London: Routledge, 1992.

[41] *Newsweek*, May 5, 1980, pp. 24–36.

[42] Nigrini, Mark J. "A Taxpayer Compliance Application of Benford's Law." *Journal of the American Taxation Association* 18 (1996), pp. 72–91.

[43] Noether, Gottfried E. *Introduction to Statistics: A Fresh Approach*. Boston: Houghton Mifflin Company, 1971.

[44] Norton, Robert M. "Pocket-Calculator Approximation for Areas Under the Standard Normal Curve." *The American Statistician* 43 (1989), pp. 24–26.

[45] Paulos, John A. *Innumeracy: Mathematical Illiteracy and its Consequences*. New York: Hill and Wang, 1988.

[46] Polya, George. *Mathematical Discovery, vol. I*. New York: John Wiley & Sons, Inc., 1962.

[47] Pringle, David. "Who's the DNA Fingerprinting Pointing At?" *New Scientist*, January 29, 1994, pp. 51–52.

[48] Rabinowitz, Stanley, ed. *Index to Mathematical Problems 1980–1984*. Westford, MA: MathPro Press, 1992.

[49] Rade, Lennart. *The Teaching of Probability and Statistics*. New York: Wiley Interscience Division, 1970.

[50] Radelet, M. "Racial Characteristics and Imposition of the Death Penalty." *American Sociological Review* 46 (1981), pp. 918–927.

[51] Remnick, David, ed. *Life Stories*. New York: Random House, 2000.

[52] Ross, Sheldon. *A First Course in Probability*. New York: Macmilllan College Publishing Company, 1994.

[53] Simpson, E. H. "The Interpretation of Interaction in Contingency Tables." *Journal of the Royal Statistical Society, Series B* 13 (1951).

[54] Sinkov, Abraham. *Elementary Cryptanalysis: A Mathematical Approach.* Washington, DC: The Mathematical Association of America, 1966.

[55] Sobel, Max A., and Evan M. Maletsky. *Teaching Mathematics: A Sourcebook of Aids, Activities, and Strategies.* Englewood Cliffs, NJ: Prentice Hall, 1988.

[56] Stern, Hal S. "A Statistician Reads the Sports Pages." *Chance* 11 (1998), pp. 46–49.

[57] Sullivan, Joseph F. "Ruling in Jersey upholds idea of equal odds for all." *New York Times*, August 13, 1985, p. B4.

[58] *Time*, April 6, 1998, pp. 29–37.

[59] *Time*, May 5, 1980, pp. 12–25.

[60] Tversky, Amos and D. Kahneman. "Causal Schemas in Judgments under Uncertainty." In *Progress in Social Psychology*, edited by M. Fishbein. Hillsdale, NJ: Erlbaum, 1980, pp. 49–72.

[61] Walbert, David. "The Effect of Jury Size on the Probability of Conviction: An evaluation of Williams vs. Florida." *Case Western Law Review* (Winter 1971): 529-55.

[62] Ward, James III. "The Probability of Election Reversal." *Mathematics Magazine* 54 (1981), pp. 256–259.

[63] Watson, Jane M. "Conditional Probability: Its Place in the Mathematics Curriculum." *The Mathematics Teacher* 88 (1995), pp. 12–17.

[64] Wetzel, Nathan. "Three Sisters Give Birth on the Same Day." *Chance* 14 (2001), pp. 23–25.

[65] Wiorkowski, John J. "A Curious Aspect of Knockout Tournaments of Size 2^m." *The American Statistician* 26 (1972), pp. 28–30.

[66] Wood, Allan James. *1918: Babe Ruth and the World Champion Boston Red Sox.* San Jose, CA: Writers Club Press, 2000.

Index